Audio Production Basics with Logic Pro X

To access online media visit:
www.halleonard.com/mylibrary

Enter Code
5179-0792-3257-1913

Audio Production Basics with Logic Pro X

Ryan Rey
Harry Gold

with contributions by

Frank D. Cook and Eric Kuehnl

NextPoint Training, Inc.
Logic Pro X

Rowman & Littlefield

Lanham • Boulder • New York • London

Published by Rowman & Littlefield
An imprint of The Rowman & Littlefield Publishing Group, Inc.
4501 Forbes Boulevard, Suite 200, Lanham, Maryland 20706
www.rowman.com

6 Tinworth Street, London SE11 5AL, United Kingdom

Copyright © 2020 by NextPoint Training, Inc.

Book design by NextPoint Training, Inc.

All rights reserved. No part of this book may be reproduced in any form or by any electronic or mechanical means, including information storage and retrieval systems, without written permission from the publisher, except by a reviewer who may quote passages in a review.

Logic Pro and all related product names and logos are registered trademarks of Apple Inc. in the US and/or other countries. All other trademarks are the property of their respective owners.

The media provided with this book and accompanying course material is to be used only to complete the exercises contained herein or for other valid educational purposes. No rights are granted to use the media in any commercial or non-commercial production.

All product features and specifications are subject to change without notice.

Library of Congress Cataloging-in-Publication Data available

ISBN 978-1-5381-3723-9 (paperback)
ISBN 978-1-5381-3724-6 (e-book)

♾™ The paper used in this publication meets the minimum requirements of American National Standard for Information Sciences—Permanence of Paper for Printed Library Materials, ANSI/NISO Z39.48-1992.

Contents

Acknowledgments xiii

Introduction. Welcome to the World of Audio xv
    Getting Started in Audio Production xv
        Who Should Read This Book xv
    About This Book xvi
        Requirements and Prerequisites xvi
        Media Files xvi
    Course Organization and Sequence xvii
    Conventions and Symbols Used in This Book xvii
        Keyboard Shortcuts and Modifiers xvii
        Icons xviii

Chapter 1. Computer Concepts 1
    Selecting a Computer 3
        Mac Versus Windows Considerations 3
        The Importance of RAM 3
        Processing Power 6
        Storage Space 7
        Storage Options 9
        Onboard Sound Options (Audio In and Out) 10
        Other Options to Consider 10
    Working with Your Computer 12
        File Management 12
        Launching Applications 17
        Saving Files 18
    Working with an Application 19
        Menus 20
        Keyboard Shortcuts 22
    Review/Discussion Questions 23

Exercise 1. Exploring Audio on the Computer 25

Chapter 2. DAW Concepts 31
Functions of a DAW 32
What Can a DAW Do? 32
Common DAWs 32
Plug-In Formats 38
What Is a Plug-In? 38
Common Plug-In Formats 39
Logic Pro X System 40
Logic Pro X Software Capabilities 40
Logic Pro X Software Options 41
Logic Pro X Hardware Options 41
Downloading and Installing Logic Pro X 41
System Requirements for Logic Pro X 42
Installation Steps 42
Launching Logic Pro X for the First Time 44
Downloadable Content 45
Important Preference Settings 45
Project Settings 47
Important Concepts in Logic Pro X 48
Project Files Versus Audio Files 48
Audio Versus MIDI 49
Review/Discussion Questions 50

Exercise 2. Setting Up a Multi-Track Project 51

Chapter 3. Audio Recording Concepts 57
The Basics of Audio 58
Frequency 58
Amplitude 59
Microphones 59
Traditional Microphones 59
USB Microphones 60
Other Considerations 61
Basic Miking Techniques 62
Multi-Tracking and Signal Flow 64
What Is Multi-Track Recording? 65
Recording Signal Flow 66

Moving Audio from Analog to Digital	66
Analog Versus Digital Audio	66
The Analog-to-Digital Conversion Process	67
The Audio Interface	68
Audio Interface Considerations	68
Working Without an Audio Interface	70
Review/Discussion Questions	71

Exercise 3. Selecting Your Audio Production Gear 73

Chapter 4. MIDI Recording Concepts 77

A Brief History of MIDI	78
Digital Control	78
The Birth of MIDI	79
The MIDI Protocol	80
MIDI Notes	80
Program Changes	81
Controller Messages	82
MIDI Controllers	83
Tone-Generating Keyboards	83
Keyboard Controllers	85
Drumpad Controllers	86
Grid Controllers	86
Alternate Controllers	87
What to Look for in a MIDI Controller	87
Purchasing a Keyboard Controller	88
Setup and Signal Flow	89
Plug-and-Play Setup	89
MIDI Cables and Jacks	89
Using a MIDI Interface	90
Considerations for Using Multiple MIDI Devices	91
MIDI Versus Audio	91
What Is MIDI?	91
Monitoring with Onboard Sound Versus Virtual Instruments	92
Tracking with Virtual Instruments	93
Creating Tracks for Virtual Instruments	93
Summary	95
Review/Discussion Questions	96

Exercise 4. Selecting Your MIDI Production Gear 98

Chapter 5. Logic Pro Concepts, Part 1 101

Creating a Project 102
- Selecting a Track Type 103
- Enabling Advanced Tools 105

Main Window 105
- The Control Bar 106
- Tracks Area Menu Bar 114
- Rulers 114

Mixer Display and Window 115

Working with Tracks 117
- Adding Tracks 117
- Mono Versus Stereo Tracks 118
- External MIDI Versus Software Instrument Tracks 118
- Regions Versus Files 119
- Selecting Tools 119

Basic Navigation 121
- Using the Transport Controls 121
- Zooming and Scrolling in the Main Window 122
- Controlling Playback Behavior 124
- Monitoring Your Playhead Location 125

Selections 126
- Selection Scenarios 128

Review/Discussion Questions 129

Exercise 5. Configuring and Working on a Project 131

Chapter 6. Logic Pro X Concepts, Part 2 137

Setting Up for Recording 138

Recording Audio 139
- Making a Record Selection 139
- Record Enabling Audio Tracks 140
- Monitoring Record Enabled Audio Tracks 140
- Initiating a Record Take 141
- Stopping a Record Take 141
- Recording Additional Takes 142

Recording MIDI 143
- Monitoring a MIDI Controller 143
- Record Enabling Software Instrument Tracks 144
- Overlap Recording on Software Instrument Tracks 145

Importing Audio and MIDI	146
Supported Audio Files	146
Importing from the Desktop	146
Importing from Media Browser and All Files Browser	146
Using the Import Command	147
Using Apple Loops	148
Working in the Tracks Area	149
Snap to Grid	149
Setting the Snap Value	149
Working with Regions and Selections	151
Selecting a Tool for Editing	151
Basic Editing Techniques	152
Advanced Editing Techniques	154
Review/Discussion Questions	158

Exercise 6. Importing and Editing Regions 160

Chapter 7. Mixing Concepts 167

Basic Mixing	168
Setting Levels	168
Panning	172
Staying Organized	175
Processing Options and Techniques	176
Gain-Based Processing	176
Time-Based Processing and Effects	177
Insert Effects Versus Send Effects	177
Processing with External Gear	177
Mixing in the Box	178
Advantages of In-the-Box Mixing	178
Getting the Most Out of an In-the-Box Mix	179
Review/Discussion Questions	180

Exercise 7. Creating a Basic Mix 182

Chapter 8. Signal Processing 189

Plug-In Basics	190
Inserting Plug-Ins on Tracks	190
Moving and Duplicating Plug-Ins	192
Displaying Plug-In Windows	192
Common Plug-In Controls	192

Adjusting Plug-In Parameters 195
    Adjusting Plug-In Parameters with the Mouse 195
    Adjusting Plug-In Parameters with the Computer Keyboard 196
EQ Processing 196
    Types of EQ 196
    Basic EQ Parameters 197
    EQ Plug-Ins in Logic Pro X 198
    Strategies for Using EQ 198
Dynamics Processing 199
    Types of Dynamics Processors 200
    Basic Dynamics Parameters 201
    Dynamics Plug-Ins in Logic Pro X 202
    Strategies for Using Compression 202
    Strategies for Using a De-Esser 203
Reverb and Delay Effects 204
    What Is Reverb? 204
    Reverb in Logic Pro X 205
    Applications for Reverb Processors 205
    What Is Delay? 206
    Delay in Logic Pro X 206
    Applications for Delay Processors 207
Wet Versus Dry Signals 208
    Using Time-Based Effects as Plug-in Inserts 208
    Using Send-and-Return Configurations 209
Review/Discussion Questions 210

Exercise 8. Optimizing Tracks with Signal Processing 212

Chapter 9. Finishing a Project 219
Recalling a Saved Mix 220
Using Automation 220
    Selecting an Automation Mode 221
    Writing Real-Time Automation 223
    Viewing Automation Curves 224
    Editing Automation Points 225
Creating a Mixdown 228
    Adding Processing on the Stereo Output 228
    Considerations for Bouncing Audio 232
    Creating a Bounce of an Audio Mix 232
Review/Discussion Questions 236

Exercise 9. Preparing the Final Mix 238

Chapter 10. Beyond the Basics 243
Flex Time and Pitch 244
Enabling Flex 244
Flex Time 245
Flex Pitch 249
MIDI Editing Techniques 251
Viewing MIDI Regions and Data 251
Editing MIDI Data 254
Working with Velocity 258
Software Instrument Region Inspector 259
Quantization in the Piano Roll Editor 260
Submixing with Track Stacks 261
Simplifying a Mix 261
Using Track Stacks to Submix 261
Creating Stems 263
Keyboard Shortcuts 264
Review/Discussion Questions 268

Exercise 10. Finalizing a Project 270

Index 277

About the Authors 285

Acknowledgments

The authors would like to give special thanks to the following individuals who have provided assistance, input, information, material, and other support for this book: Eric Kuehnl, Frank Cook, Jules Fuchs.

Harry would like to thank his family, and musical mentors Steve Gannon and Michael Burks.

Ryan would like to thank his wife Audrey for her support and understanding throughout the many late nights; his daughter Olivia for naps that lasted longer than usual so work could continue during the day; his former mentor Hdez for nurturing a deeper passion for technology and education; and finally his Grandma Shirley for her encouragement and enthusiasm for all the projects shared with her. She will be missed.

Introduction

Welcome to the World of Audio

Sounds are all around us. They make our world interesting, informative, and engaging. It's only natural to want to capture these auditory experiences—the sounds we like, the sounds we want to share, and the sounds we create. That is truly what audio production in Logic Pro X is all about: creating, capturing, and sharing sound.

Logic Pro X provides a robust platform for learning the fundamentals of audio production and improving the results of all your audio endeavors. This book has been written for readers who are new to Logic Pro X; however, the features discussed in these pages provide plenty of depth for experienced users as well. The included exercises can be completed with a basic Logic Pro X installation. No additional hardware or software is required.

With this book, you're taking the first step toward discovering the unique capabilities of Logic Pro X software and unlocking your own audio creativity. Whether you've selected this book to use for self-study or have picked it up as the required text for instructor-led classroom training, you will find that it covers the principles of audio production from the ground up. It also gives you everything you need to know to fully understand the role of Logic Pro X in today's landscape of digital audio workstations (DAWs).

Getting Started in Audio Production

This book teaches the basics of recording, editing, mixing, and processing audio and MIDI using Logic software. It also provides plenty of power tips to take you beyond the basics and unleash the true power of using Logic Pro X as a creative tool. Additionally, the principles you learn in this book will apply equally to other commercial products used to create, record, edit, and process digital audio. This means you'll have a solid foundation, if you should supplement your work in Logic with another digital audio workstation in the future, or even find yourself working in a full-fledged commercial studio down the line.

Who Should Read This Book

Although this book is written to support beginners working with Logic Pro X software, the concepts apply equally to professional work on very large projects. As such, this book serves as a resource for users at any level. Learners who have been using Logic Pro X for years are bound to discover nuances in the software that they may have been unaware of or have not had the opportunity to explore.

This book is designed to help new and inexperienced users get started with little or no background knowledge. However, the scope is not limited to novice users who are experimenting with DAWs for the first time.

About This Book

This book is designed for use in a formal course of study. The text and the associated instructor-led course were developed by NextPoint Training, Inc. (NPT) as part of our Digital Media Production program and certification offerings. While this coursebook can be completed through self-study, we recommend the hands-on experience available through an instructor-led class with an NPT Certification Partner.

For more information on the classes offered through the NextPoint Training Digital Media Production program, please visit https://nxpt.us/DigitalMedia. For information on how your school can become an NPT Certification Partner, please visit https://nxpt.us/CertPartner.

Requirements and Prerequisites

This course does not require any specific background knowledge of computer systems, recording technology, or digital audio workstations. However, before starting to work with Logic Pro X, it definitely helps to have at least a passing familiarity with computer concepts and recording gear. If you consider yourself a novice in these areas, pay special attention to the first four chapters of this book.

To try out the concepts and complete the exercises in this book, you will need to use a compatible computer and install Logic Pro X software along with its associated audio resources. Details on computer requirements are provided in Chapter 1, and installation details for Logic Pro X are provided in Chapter 2.

If you will be using other connected audio or MIDI hardware, you may need to install device drivers for those components. Consult the documentation that came with your hardware or search the manufacturer's website for details.

Media Files

This book includes exercises at the end of each chapter that make use of various media files. The media files can be accessed by visiting www.halleonard.com/mylibrary and entering your access code, as printed on the opening page of this book. Instructions for downloading the media files are provided in Exercise 1.

The media files for the exercises in this book have been provided courtesy of The Pinder Brothers, Eric Kuehnl, and Fotograf.

Course Organization and Sequence

This course has been designed to teach you how to get the most out of your work with Logic Pro X software. The material is organized into 10 chapters, as follows:

- **Chapter 1. Computer Concepts**—What you need from a computer
- **Chapter 2. DAW Concepts**—What you need from your DAW
- **Chapter 3. Audio Recording Concepts**—What you need to record audio
- **Chapter 4. MIDI Recording Concepts**—What you need to record MIDI
- **Chapter 5. Logic Pro X Concepts, Part 1**—What you need to know to get started with Logic Pro X software
- **Chapter 6. Logic Pro X Concepts, Part 2**—What you need to know to work with Logic Pro X software
- **Chapter 7. Mixing Concepts**—What you need to know to mix a project
- **Chapter 8. Signal Processing**—What you can do to optimize your audio
- **Chapter 9. Finishing a Project**—What you need to do to create a stereo mixdown
- **Chapter 10. Beyond the Basics**—What to explore to become a power user

Users who have experience with computers and other DAWs may wish to skim the first four chapters, focusing mostly on the details that are specific to Logic Pro X software.

Conventions and Symbols Used in This Book

Following are some of the conventions and symbols we've used in this book. We try to use familiar conventions and symbols whose meanings are self-evident.

Keyboard Shortcuts and Modifiers

Menu choices and keyboard commands are typically capitalized and written in bold text. Hierarchy is shown using the greater than symbol (>), keystroke combinations use the plus sign (+), and mouse-click operations use hyphenated strings, where needed. Brackets ([ ]) indicate key presses on the numeric keypad.

Convention	Action
File > Save Session	Choose Save Session from the File menu.
Ctrl+N	Hold down the Ctrl key and press the N key.
Command-click (Mac)	Hold down the Command key and click the mouse button.
Right-click	Click with the right mouse button.
Press [1]	Press 1 on the numeric keypad.

Icons

The following icons are used in this book to call attention to tips, shortcuts, listening suggestions, warnings, and reference sources.

Tips provide helpful hints and suggestions, background information, or details on related operations or concepts.

Shortcuts provide useful keyboard, mouse, or modifier-based shortcuts that can help you work more efficiently.

Power Tips provide shortcuts and tips for power users that can dramatically speed up your work but that go beyond the scope of the current discussion.

Listening Suggestions refer you to audio examples that illustrate a concept or technique discussed in the text.

Warnings caution you against conditions that may affect audio playback, impact system performance, alter data files, or interrupt hardware connections.

Cross-References alert you to another section, book, or resource that provides additional information on the current topic.

Online References provide links to online resources and downloads related to the current topic.

Chapter 1

Computer Concepts

...What You Need from a Computer...

In this chapter, we introduce you to the basic components of a computer system and the minimum configuration required to run Logic Pro software. We also discuss how to configure the hardware for your system and how to configure your computer's software settings for optimal results. Lastly, we take a brief look at how to work within your digital audio workstation, using Logic Pro X as an example.

⊕ Learning Targets for This Chapter

- Understand how to select a computer for your digital audio workstation
- Understand how to navigate your computer's operating system for basic file management purposes
- Understand how to set preferences for your computer and your software
- Understand how to perform basic operations with your digital audio workstation

Key topics from this chapter are illustrated in the Logic Audio Production Basics Study Guide module available through the Elements|ED online learning platform. Sign up at ElementsED.com.

Selecting the right digital audio workstation (or DAW) involves many considerations: What features do you need? How big will your projects be? What kind of production work do you intend to do? What kind of system can you afford? And so on. Prior to making a purchase, you would be wise to check out a trial version or low-cost (feature-limited) edition of the product you are considering, if available.

For users familiar with GarageBand, Logic Pro may be a compelling option for a full-featured DAW.

> ### About GarageBand
> Apple offers GarageBand software completely free of charge. Although GarageBand does have certain limitations, it can be a good option to get started working with audio and MIDI. GarageBand provides a streamlined user experience and a gentler learning curve than Logic Pro, making it a good choice for new users. Using GarageBand will also introduce you to various functions and conventions that carry over to Logic Pro, easing the transition should you decide to upgrade.

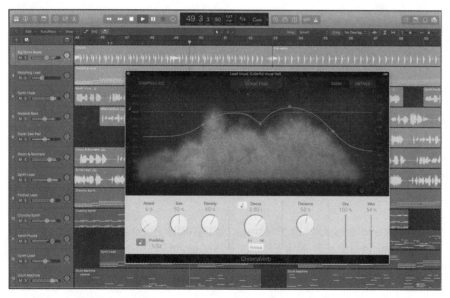

Figure 1.1 Logic Pro X software with the Chromaverb plug-in window open

Before installing Logic Pro (or other DAW software), you'll need to verify that your computer system meets the minimum requirements for the software. Once you've cleared that hurdle and installed the software on a suitable computer, you will want to configure your computer and the software application for optimal performance. This chapter will help steer you in the right direction for these and other considerations to get a system up and running successfully.

Selecting a Computer

In this section, we will look at how to select a computer for use with Logic Pro X or another digital audio workstation of your choice. We will also look at options for optimizing your computer setup to get the most out of your work environment.

Mac Versus Windows Considerations

One of your first decisions is whether to run your DAW on a Mac or a Windows computer. This choice is primarily one of personal preference, but it may also be based on the kind of system you already own.

If you are purchasing a new system for your DAW, understanding the differences between Mac computers and Windows machines may help you decide on one or the other. In some cases, your options may be limited by support for your DAW of choice. Many DAWs run on both Mac- and Windows-based computers. However, Apple's Garage Band and Logic Pro X are exclusively Mac-based.

Mac-Based Computers

One of the big selling points for Macs is how well they interact with other Apple products: iPhones, iPads, Apple TVs, and the Apple Watch. If you have already bought in to the Apple ecosystem, it may make sense to purchase a Mac computer for that reason alone. You will find the user experience on a Mac to be similar to that on other Apple products, in many respects.

Mac systems also have a reputation for being easy to use and simple to understand. Macs are known for high quality construction and attention to detail in their design.

On the other hand, you will likely spend more for a Mac-based system than you would for a similarly configured Windows-based system.

Windows-Based Computers

Windows-based computers are available from many different manufacturers, giving you an abundance of choices when selecting a system. The competition between manufacturers can make it easier to find a system that meets your needs at a price point you can afford.

Windows computers can also have an advantage for certain applications. If you use Windows-only applications for personal or professional purposes, that may influence your choice of platform.

Whichever platform you choose—Mac or Windows—be sure to purchase a system with adequate RAM, processing power, storage space, and connectivity options to support your needs.

The Importance of RAM

A computer's installed RAM, or random access memory, determines how much data and information can be stored in the computer's memory at any given time. RAM is generally used for temporary storage while an application is running. The RAM allocation is typically measured in Gigabytes (or GB).

Figure 1.2 Illustration of a typical RAM module

> ### Running Windows on a Mac
>
> If you prefer to work on a Mac but need to run Windows for certain applications, several options are available. Modern Mac computers allow you to run a Windows operating system (OS) on your Mac hardware. You can configure a Mac to run Windows 10, for example, using either a separate disk partition or a virtualization software option.
>
> To configure a separate disk partition for Windows 10, you can use the Boot Camp Assistant utility that comes with your Mac. This utility will partition your drive and guide you through the installation process for the Windows software. Using this option, you will need to restart your computer whenever you wish to switch between the Mac OS and the Windows OS.
>
> To run Windows 10 with virtualization software instead, you can purchase and install software such as VMware Fusion or Parallels Desktop. These applications allow you to run a Windows OS on your Mac desktop without rebooting.
>
> With either option, you will need to purchase Windows 10 separately and install it. Factor this in when considering the total budget for your system. The virtualization software and the Windows OS may also impact the amount of RAM, storage space, and processing power required for the system.

How RAM Is Used

When you select a paragraph of text in a word processor and choose the **EDIT > COPY** command, the text is stored temporarily in RAM. You can then move your cursor to a new location in the document and choose **EDIT > PASTE**. The text stored in RAM gets added to the document at the current cursor location.

Most applications also use RAM to keep track of the work you do in your open documents. As you work, a record of your changes and edits is saved in RAM. This allows you to use the **EDIT > UNDO** command, for example, to reverse any changes you've made.

The code required to run the application is also stored in RAM while the application is running. This allows the computer to work faster than it would if it had to run the application entirely from the hard drive.

Storing data in RAM allows the information to be retrieved and operated on very quickly.

However, once you quit an application or shut down your computer, the information stored in RAM is lost. That is why your changes must be saved to permanent storage first.

Why More RAM Is Better

DAWs such as Logic Pro X rely on RAM to work efficiently. Like a word processor, a DAW uses RAM for edit operations. Your DAW will also use RAM to keep track of your changes and undoable actions.

 In Logic Pro X software, we call the list of user-undoable actions the *Undo History*.

The more RAM your computer system has installed, the more memory will be available to your DAW software for temporary storage. This means the application can run more efficiently, you can copy and paste larger amounts of data, and you can maintain a longer list of complex undoable actions.

Symptoms of insufficient RAM can include sluggish behavior, frequent error messages, long pauses (or hangs) when you perform certain actions, and sudden crashes.

DAW manufacturers typically publish minimum RAM specifications required to run the application. They may also suggest that you install more RAM for more efficient operation. For work on very large or complicated projects, you may want to exceed the manufacturer recommendations.

RAM Required for Logic Pro

The minimum RAM allocation to run Logic Pro X on a Mac computer is 4 GB. However, you may want to consider increasing that to 8 GB or more. Most other professional DAWs require a minimum of 16 GB. It's a good idea to get as much as you can. If you plan to run Logic Pro X at the same time as other applications (iTunes, a web browser, an email client, etc.), you should target the high side of these numbers.

> **Installing More RAM**
>
> Depending on the model computer you buy, you may be able to add more RAM at a later time, as your needs grow. This can help keep the initial purchase price down. If this is your plan, make sure to do some research before purchasing a computer. Certain models (especially budget options) have RAM chips soldered in place and cannot be upgraded later with more. Many Apple laptops such as the MacBook Pro do not have user upgradable RAM, so you'll want to plan out how much you need in advance. Smaller desktop systems may allow you to add RAM later.
>
> Also note that upgrading RAM often requires you to replace all RAM in the computer, rather than simply adding to the existing RAM. Most computers have an even number of memory slots, with RAM assigned in pairs across them.
>
> For example, a laptop may have 4 GB of RAM across two available memory slots. In this case, each slot will have a 2 GB memory module installed. To upgrade to 8 GB of RAM, you would need to replace each module with a new 4 GB module. (RAM modules should be added in matching pairs.)

Processing Power

The processing power, or how fast a computer runs, is a function of the computer's central processing unit, or CPU. The CPU is responsible for performing all of the processing and calculations required by an application.

Figure 1.3 The computer's central processing unit determines how fast the computer can run.

Processing Speed. The speed of a CPU is based on its clock speed. Clock speeds for modern computers are measured in GHz (Gigahertz), or billions of cycles per second. A 1 GHz processor, for example, can process up to one billion instructions per second. The faster the CPU, the more powerful the computer, and the faster your applications will run.

Processor Cores. Another consideration is the number of processing cores in the computer chip. Many computer chips today offer dual cores, quad cores, or more. This means that the CPU includes multiple processing units within a single chip. The benefit of multiple cores is that the computer can process multiple instructions at one time. This in turn allows modern applications to run faster and more efficiently.

> ### Hyper-Threading
> Many CPUs support hyper-threading as an alternative (or supplement) to multiple physical cores. Intel's hyper-threading technology allows a single physical core to perform as two virtual cores. This has the same benefit as adding physical cores, as it doubles the number of instruction threads that the CPU can process at once when running an application.

How Processing Power Affects Your DAW

A host-based DAW like Logic Pro X relies on the processing power of the host computer for all audio editing, mixing, and processing. A system with insufficient processing power may exhibit sluggish performance, unwanted audio artifacts, and frequent error messages. Projects involving large track counts and heavy use of plug-ins will be more susceptible to issues.

Certain audio processes are especially CPU-intensive. Some examples of common practices that may tax your CPU include the following:

- Using certain Edit operations that automatically split audio into many smaller pieces
- Stacking plug-ins on inserts, especially convolution reverbs and analog gear emulations
- Using many virtual instrument plug-ins in a project

- Using real-time Flex-Time processing
- Configuring complex signal routing for submix groups, headphone outputs, and effects processing

To avoid running into processing limitations as your projects grow in complexity, look for a multi-core CPU that runs at a reasonably high speed.

Processing Requirements for Logic Pro X

To run Logic Pro X, you need to be able to run macOS 10.13.6 or later. This means the oldest OS you will be able to use to run Logic Pro X is Apple's *High Sierra*.

You also need an OpenCL-capable graphics card or Intel HD Graphics 3000 or later. This should have a minimum of 256mb of VRAM (video RAM).

Most Apple computer systems available today meet the minimum CPU requirements to run Logic Pro X.

 Logic Pro X supports multithreading, so you will get better performance out of a quad-core processor than you will out of a duo-core, for example.

Always Check Compatibility Before You Buy!

New versions of Logic Pro X often have higher requirements than previous versions. Prior to buying a system, be sure to check the compatibility requirements on the Apple website.

Current requirements for Logic Pro X can be found online at: https://www.apple.com/logic-pro/specs/

Storage Space

Another important consideration when buying a computer is the amount of storage space the computer provides. This is a function of the computer's installed hard disk drive (HDD) or solid state drive (SSD). The size of the storage drive determines how many applications you can install and how much space you will have for saving files on your system.

Figure 1.4 Illustration of a hard disk drive

Determining How Much Drive Space You Need

When determining storage requirements for your DAW, you will need to consider multiple factors:

- How much space is required to install the DAW software application itself.
- How much space is required to install the plug-ins you intend to use with the DAW.
- How much space you will need for the projects and media files that you create with the DAW.
- How much space you will need for other media files that you may want to use with the DAW (such as sample libraries, loop libraries, and sound effects).

In addition, you may need to consider the impact of other files and applications that you want to use on your computer. For example, if you plan to store your digital photos and your personal music collection on the computer, you will need to allocate additional space for those purposes.

Likewise, if you will use the computer for work in programs such as Microsoft Word, PowerPoint, and Excel, you will need additional space to install each of those applications.

Installation Requirements for Logic Pro X

Installing Logic Pro X software requires a minimum of 6 GB of disk space, for the software alone, and up to 63 GB for a full installation.

 Logic Pro X comes with a large library of virtual instruments and loops. You can install all of this or pick and choose which libraries you want on your machine. A full installation takes up approximately 63 GB of storage space.

Storage Options

Many storage options are available for today's computer systems. A computer's internal system drive may be either an HDD or an SSD. The internal storage can be supplemented by any number of external drives and storage options. To allow your files to be accessed from anywhere, you can also consider cloud-based storage.

HDD Versus SSD Storage

HDD. Hard disk drives use spinning metal disks to store data. The more disks in the drive, the higher the data capacity. Storage sizes for modern HDDs typically range from around 1 to 4 Terabytes (TB).

The data transfer speed for an HDD is based in part on the speed of the spinning disks, measured in revolutions per minute (or RPM). Common options include 5400 RPM drives and 7200 RPM drives.

Hard disk drives provide greater storage capacity at a lower price than their solid-state counterparts.

SSD. Solid state drives, by comparison, use flash memory for storage instead of spinning metal disks. The storage capacity of an SSD is based on the number and size of memory chips it has. Storage sizes for SSDs typically range from 256 GB to 1 TB.

Solid state drives provide faster access speeds than HDDs, meaning they provide better read/write performance for the computer's functions. This is noticeable in faster start-up times for the computer, faster launch times for applications, and faster file transfers among systems and drives.

Solid state drives also require less power than HDDs. This gives them an advantage for laptop computers and mobile devices. Additionally, SSDs have no moving parts, so they operate silently and are less prone to mechanical failure. These advantages come at the expense of higher cost and lower overall storage capacity.

External Drive Storage

A popular option for supplementing a computer's storage is to use an additional, external drive. External drives are readily available, with a variety of connection types. Some common types include the following:

- FireWire 400 or 800
- USB 2.0 or 3.0
- USB-C
- eSATA
- Thunderbolt 1, 2, or 3

If you plan to store media files on an external drive for use with your DAW, look for a high-performance drive with a high-speed data connection to your computer. Good drive options would include a 7200 RPM HDD or an SSD. Good data connections would include USB 3.0, USB-C, eSATA, Thunderbolt 2, or Thunderbolt 3. Be sure to match the capabilities of your computer when selecting the connection type.

Onboard Sound Options (Audio In and Out)

Another aspect to consider when selecting a computer is the onboard sound options that the model offers. Many DAWs, including Logic Pro X, can utilize the computer's built-in sound options for recording and playback. This allows you to use any built-in microphone inputs on the computer for recording to your project and to use the built-in computer speakers for playing back your project.

While you may not want to use the computer's microphones for any meaningful recording project, having onboard audio will allow you to run your DAW without requiring a connected audio interface. This can save you some money at the outset. It also provides the flexibility to work on the go, especially when using a laptop, without needing to bring along additional hardware.

 Although audio inputs are not required, audio outputs are critical. Look for a computer that has built-in speakers and/or a stereo headphone jack for monitoring playback from your DAW.

Other Options to Consider

Aside from the computer hardware itself, you may also want to consider options for peripheral devices. With the right accessories, you can customize your setup for efficiency and/or personal preference.

Mice and Trackballs

Many computer users find the mouse that comes with their computer to be less than ideal for long-term use. Numerous alternatives are available, offering features such as multiple buttons, scroll wheels, trackballs, gesture/touch input, and wireless connection to the computer. Mouse upgrades can range from around $20 to upwards of $100. Be sure to try out any options you are considering while running your DAW (or a similar application) to determine what works best for you.

Trackpads and Touchpads

As an alternative or supplement to a traditional mouse, Apple offers several Trackpad options (such as the Magic Trackpad and Magic Trackpad 2). These devices provide touch-based gesture input, similar to that found on many modern laptops. Logitech, Dell, and others offer similar Touchpad devices.

Extended Keyboards and Keypad Options

An upgrade option that is well worth the investment is an extended keyboard. If you plan to do any serious work in Logic Pro X, an extended keyboard is a must. The extended keyboard provides access to numeric keypad keys on the right-hand side, which are used extensively in Logic Pro X.

Figure 1.5 Standard Mac keyboard (top) and extended Mac keyboard (bottom)

If you own or purchase a laptop computer, you can opt for a numeric keypad extension instead. These are available from various manufacturers and can be connected to your laptop wirelessly or via USB cable, depending on the model.

Figure 1.6 Numeric keypad extension option from Cateck

Benefits of an Audio Interface

Many audio interfaces are available that are compatible with Logic Pro X. Any audio interface that is Core Audio-compliant will work with Logic Pro X.

Some good options include the **Focusrite Scarlett** series or **Focusrite iTrack Solo**. These Focusrite products each come bundled with several plug-ins from companies like XLN and Softube. (See https://focusrite.com/usb-audio-interface/scarlett/scarlett-2i2.) Many Focusrite interfaces are also available in **Studio** editions, with microphone and headphones included.

> The benefits of using an audio interface can include better sound quality, a greater number of available inputs and outputs (I/O), and access to digital I/O options. For many beginning users, the number of inputs may be the most significant issue.

Working with Your Computer

Once you have selected an appropriate computer to host the DAW of your choice, you should spend some time to get familiar with basic operations on the system. If you already have experience with computers and the operating system you will be using, you may want to skip or skim this section. The information in this section focuses on navigating among the files, folders, and applications on your computer.

In this section, we cover basic operations such as locating, moving, and organizing files; creating folders; launching applications; and saving files.

File Management

Perhaps the most important thing you should know about your computer is how to navigate among the files and folders using the computer's operating system. On the Mac, you will use the Finder for this purpose.

Using the Finder (Mac-based OS)

The desktop experience in a Mac-based operating system is defined by the Finder application. The Finder is the first thing you see after you start up a Mac, before you launch any applications or open any files or folders.

Figure 1.7 The Finder icon on a Mac

The Finder includes a menu bar at the top of the screen, the desktop display in the center (background image along with any desktop icons and open Finder windows), and the Dock at the bottom of the screen. The term "Finder" is commonly used interchangeably with the term "desktop" on a Mac.

Computer Concepts 13

Figure 1.8 The desktop in the Mac Finder (OS X 10.11 El Capitan)

The Finder Versus the Desktop

Technically speaking, the Finder and the desktop are two different things. The Finder is an application on the Mac that is always running. You can switch to the Finder application at any time, the same way you can switch between other running applications.

The desktop is the default location displayed by the Finder. You can think of the desktop as the background canvas upon which your open folders and applications are displayed. The desktop also provides a location to store folders and files.

The items stored on the desktop are visible from the Finder. They can also be displayed in a Finder window, by opening the Desktop folder.

From the Finder, you can:

- Navigate through the file system on your storage drive(s)
- Open folder locations, and
- Manage files within open folders

To open a new Finder window, choose **FILE > NEW FINDER WINDOW** from the menu bar at the top of the screen (or double-click on a folder on the desktop). You can navigate within an open Finder window by clicking on a location in the sidebar or by double-clicking on a displayed folder in the file list to open it.

Figure 1.9 An open Finder window (Mac OS X)

Using File Explorer (Windows-based OS)

Windows computers provide a desktop view similar to the Mac Finder. However, the desktop might not be the first thing a user sees upon startup, depending on the operating system. (Windows 8 and Windows 8.1 use the "Metro UI" tile view by default, similar to that on a Windows Phone.)

To open a new File Explorer window, do one of the following:

- In Windows 7 or Windows 10, click on the yellow File Explorer icon in the taskbar at the bottom of the desktop screen (or double-click any folder on the desktop).

Figure 1.10 Mouse cursor next to the File Explorer icon in the taskbar (Windows 7 shown)

- In Windows 8 or Windows 8.1, first click the **DESKTOP** tile from the tiled Start screen; then from the desktop, click the File Explorer icon in the taskbar (or double-click any folder on the desktop).

> In Windows 8.1, you can also access the File Explorer directly from the tiled Start screen by typing *file explorer* and clicking on the File Explorer icon that appears.

As in the Mac Finder, you can navigate within an open File Explorer window by clicking on a location in the Navigation pane on the left or by double-clicking on a displayed folder in the file list to open it.

Locating, Moving, and Opening Files

Using the locations in the sidebar of a Mac Finder window (or the Navigation pane in File Explorer), you can navigate to different folder locations on your computer. For example, clicking on the **DOWNLOADS** location will take you to the Downloads folder on your system (for any files you've downloaded from the Internet). Clicking on the **DOCUMENTS** location will take you to your Documents folder.

You can access the files within a folder at a given location by double-clicking on the folder to open it. On a Mac, you can also click on the triangle next to a folder to expand it and display its contents.

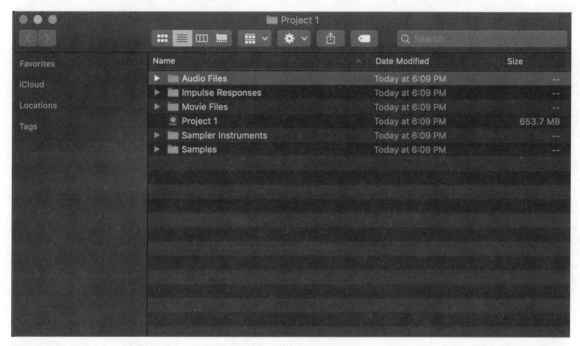

Figure 1.11 Finder window with a Logic Pro X project folder displayed

Moving Files. To move a file from one location to another, you can drag the file from its current location to any other visible folder within the window or to any location displayed in the sidebar/Navigation pane. You also can drag a file from one open window to another.

Opening Files. Opening a file is usually as simple as double-clicking on the file. In most cases, the file will open in the application that created it, if available on your computer. If the application is not available—or if your computer cannot determine an appropriate application to use for the file—you can choose a compatible application on your system. On both Mac- and Windows-based systems, you can right-click on a file and select an application to use for the file from the right-click menu.

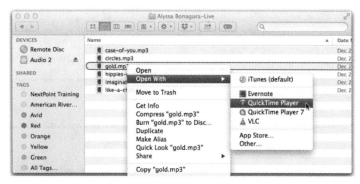

Figure 1.12 Right-click menu for a file in a Mac Finder window

> ### Accessing Right-Click Functions on a Mac
>
> For users who are new to the Mac, using the mouse can require an adjustment. The Apple mouse doesn't offer separate left- and right-click buttons. Instead, you must rock the mouse to get the desired effect. For a standard left-click, you must click on the *left side* of the mouse, causing the mouse to rock left. For a right-click, you need to click on the *right side*, causing it to rock right.
>
> To improve your results, practice using two fingers on the mouse. Place your index finger on the left side and your middle finger on the right. (Reverse if you are left-handed.) For a left-click, roll your hand to the left and click with your index finger. For a right-click, roll to the right and click with your middle finger.
>
> If you are using a Mac trackpad, use a two-finger click to access right-click functions.

Creating Folders

On both Windows and Mac, you can easily create a new folder on the desktop or at another desired location. While displaying the desktop or viewing a location in a Finder or File Explorer window, simply right-click with the mouse and select **NEW FOLDER** (Mac) or choose **NEW > FOLDER** (Windows) using the pop-up menu. (See Figures 1.13 and 1.14.)

You can also create a new folder by using a menu command (Mac) or Ribbon button (Windows 8 and later) or by using associated keyboard shortcuts: **COMMAND+SHIFT+N** (Mac) or **CTRL+SHIFT+N** (Windows).

Computer Concepts 17

Figure 1.13 Creating a folder by right-clicking on a Mac

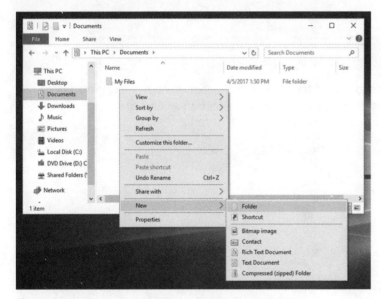

Figure 1.14 Creating a folder by right-clicking on Windows

Launching Applications

To launch an installed application, you can either double-click on the app file in your computer's Applications directory or use one of the shortcut operations provided by your OS.

Mac-Based Systems. The easiest path to opening an application on the Mac is the Dock. The Dock is a bar of icons that sits at the bottom of your Mac screen by default. It provides quick access to your installed apps. To launch an app, simply click on its icon in the Dock.

Figure 1.15 Launching Logic Pro X from the Dock

Windows-Based Systems. One easy way to launch an application on any Windows system is to use the Search function. Simply press the **START** key (also known as the **WINDOWS** key) and begin typing the name of the application you wish to open.

 Pressing the START key followed by any text activates a Start Menu search in Windows 7, a search in a right-hand pane in Windows 8, or a Cortana search in Windows 10.

Launching an App from the Desktop

During the install process, many applications will offer the option to place a shortcut icon, or *alias*, on your desktop. This shortcut links to the application file in your Applications directory.

Figure 1.16 Logic Pro X shortcut on the Mac Desktop

Shortcut icons provide a convenient option for launching an application, as you can simply double-click on the shortcut from the desktop to start the application.

Saving Files

Most applications allow you to start working on a file and save it at a later point, once you've created something worth keeping. Logic Pro X works this way as well, storing your work in a temporary location until you decide to save the project. The usual process for saving your work is to choose **FILE > SAVE** using the menus at the top of the application.

When you save a Logic Pro X project for the first time, you can choose a name and save location for your project. Logic Pro X saves your work using a hierarchy of files and folders.

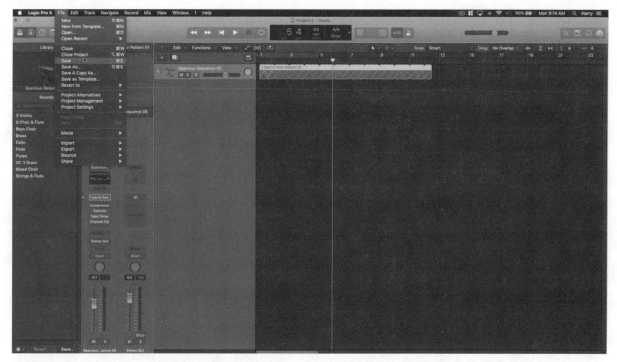

Figure 1.17 Saving using the File menu in Logic Pro X

> ### The Logic Pro X File Hierarchy
>
> The concept of a file hierarchy is common to all varieties of Logic Pro X software. This structure is used to allow the software to reference media files, rather than embedding them inside of the Logic Pro X file itself.
>
> For example, any audio files that you record or import into Logic Pro X will be stored separately within the project hierarchy. The actual Logic Pro X file simply links to the audio files, keeping the size of the Logic Pro X file itself to a minimum.

Once you have created your initial project, you can save the changes you make as you work by using the standard FILE > SAVE command.

Working with an Application

With your application up and running, you will be able to begin working on a project—whether that be a slide presentation, a spreadsheet document, or an audio recording project. Most modern applications, including Logic Pro X, use menus and keyboard shortcuts for accessing their main commands.

Menus

The term *menu* is commonly used in software to denote one of the pull-down lists of commands at the top of the application. On Windows computers, the menu bar is typically anchored to the top of the application window. However, on Mac computers, the menu bar is typically anchored to the top of the screen.

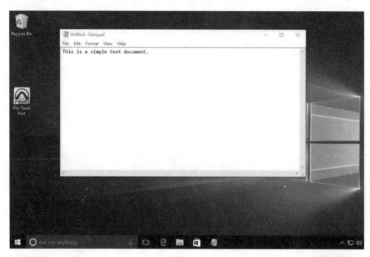

Figure 1.18 Menus in the Notepad app on Windows 10 (top of application window)

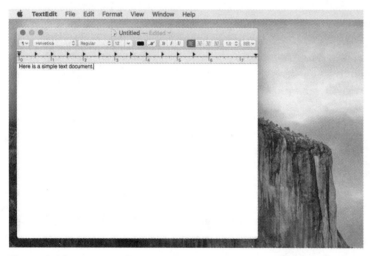

Figure 1.19 Menus for the TextEdit app on Mac OS X 10.11 (top of the screen)

To select a command, click on the associated menu name in the menu bar (such as the **FILE** menu) and then click on the desired command from the pull-down menu (such as the **SAVE** command). In some cases, menu items will include a submenu that you can use to select from between multiple choices.

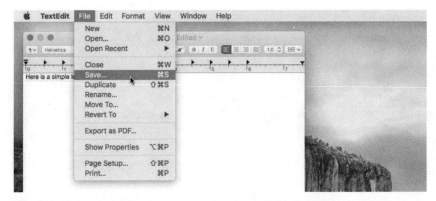

Figure 1.20 Selecting the Save command under the File menu

At times, you may have to distinguish between menus used by different applications or by the operating system (such as the Mac Finder). On a Windows-based system, you will likely find it easy to tell which menus belong to a given application, since the menus are anchored to the application window.

On a Mac-based system, however, you can run into situations where an application's windows are closed, but the application is still active. In these cases, you may see the desktop or another running application on screen, making it less apparent which application is active.

Figure 1.21 Logic Pro X menus displayed on a Mac with all application windows closed

To determine which application is active on a Mac, look for the application menu on the left side of the menu bar (immediately to the right of the Apple menu). The name of the active application will always display, regardless of whether the application has any windows open.

 Clicking on the desktop screen, an open Finder window, or a background app will make that item active, replacing the menus at the top of the screen.

 You can toggle through open applications by pressing COMMAND+TAB on a Mac or ALT+TAB on Windows.

Keyboard Shortcuts

As an alternative to using the menus to access commands, you may also be able to use keyboard shortcuts or other key commands. Keyboard shortcuts use one or more modifier keys in combination with a standard key press to activate a command. Common examples include **COMMAND+S** on the Mac (**CTRL+S** on Windows) to save the open file or **COMMAND+C** on the Mac (**CTRL+C** on Windows) to copy a selection.

Mac and Windows systems each use different modifier keys, but their functions are similar. Table 1.1 shows the modifier keys available on each platform.

Table 1.1 Modifier keys available on Mac and Windows systems

Modifier Keys on Mac-based Computers	Modifier Keys on Windows-based Computers
The **COMMAND** key (⌘)	The **CTRL** key
The **OPTION** key (⌥)	The **ALT** key
The **CONTROL** key (^)	The **START** key (⊞)
The **SHIFT** key (⇧)	The **SHIFT** key

 As a Mac-only DAW, Logic Pro X uses only the Mac modifier keys. The remainder of this book will focus primarily on Logic software and Mac-based operations.

The advantages of using keyboard shortcuts include speed and efficiency, as well as reduced dependency on the mouse. This not only makes your work more precise, but it also helps avoid ergonomic problems, including carpal tunnel syndrome.

Often, keyboard shortcuts are listed next to the command names in the pull-down menus. You'll find it worth your time to memorize the shortcuts for some of the commands you use frequently in your DAW. In addition to the shortcuts shown in menus, Logic Pro X includes numerous other "hidden" keyboard shortcuts. We will be covering many of these in this book.

Review/Discussion Questions

1. What are some reasons for choosing a Mac computer for your DAW? What are some reasons for choosing a Windows computer instead? (See "Mac Versus Windows Considerations" beginning on page 3.)

2. How is RAM used by an application? Is RAM used for permanent storage or temporary storage? (See "How RAM Is Used" beginning on page 4.)

3. What are some indications that a computer system does not have enough installed RAM for the applications you are using? (See "Why More RAM Is Better" beginning on page 5.)

4. What factors affect how much processing power your computer has for running applications? (See "Processing Power" beginning on page 6.)

5. What are some examples of audio processes that can strain the CPU of an underpowered system? (See "How Processing Power Affects Your DAW" beginning on page 6.)

6. What are some things you should consider to determine how much storage space is required for your DAW computer system? (See "Determining How Much Drive Space You Need" beginning on page 8.)

7. What is the difference between an HDD and an SSD for storage? What are some advantages of each? (See "Storage Options" beginning on page 9.)

8. How are built-in sound options on a computer useful for Logic Pro X? (See "Onboard Sound Options" beginning on page 10.)

9. Why might you want to consider an extended keyboard for your Logic Pro X system? (See "Extended Keyboards and Keypad Options" beginning on page 10.)

Chapter 1

10. How can you quickly navigate to different folder locations on your computer? (See "Locating, Moving, and Opening Files" beginning on page 15.)

11. What are some ways to create a new folder at the current location on your computer? (See "Creating Folders" beginning on page 16.)

12. How can you select a command in an application such as Logic Pro X? (See "Menus" beginning on page 20.)

13. What modifier keys are available on Mac computers? What modifiers keys are available on Windows computers? (See "Keyboard Shortcuts" beginning on page 22.)

 To review additional material from this chapter and prepare for certification, see the Logic Audio Production Basics Study Guide module available through the Elements|ED online learning platform at ElementsED.com.

Exercise 1

Exploring Audio on the Computer

🎧 Activity

In this exercise, you will download the media files used for this course and move them to an appropriate storage location on your system. Then you will use your computer to listen to sample files for an excerpt from the song "Overboard" by Sacramento-area band, The Pinder Brothers.

🕓 Duration

This exercise should take approximately 10 minutes to complete.

⊕ Goals/Targets

- Practice file management techniques
- Practice critical listening techniques
- Explore sonic elements that comprise a finished mix

Exercise Media

This exercise uses media files taken from the song, "Overboard," provided courtesy of the band The Pinder Brothers.

Written by: Matt Pinder; Performed by: The Pinder Brothers;
Produced by: Scott Reams and The Pinder Brothers*

The media provided for this course may be used for educational purposes only. No rights are granted to use the media for any other personal, commercial, or non-commercial purposes.

* The mix, processing, and media files have been adapted for use in the exercises contained herein.

Getting Started

To get started, you will download the media files for this course and locate them in your **Downloads** folder. You will then move the downloaded media files to your **Documents** folder or other appropriate location.

Download the course media files:

1. Launch the browser of your choice on your computer (Safari, Firefox, Chrome, etc.)

2. Point your browser to **www.halleonard.com/mylibrary** and enter your access code (printed on the opening page of this book).

3. Click the **DOWNLOAD** link next to the **Logic APB Media Files** listing in your **My Library** page. The **Logic ABP Media Files** folder will begin downloading to your Downloads folder.

4. Open the Downloads folder on your system by activating the Finder and choosing **GO > DOWNLOADS**.

Move the Logic ABP Media Files folder to a permanent location:

1. Select the **Logic ABP Media Files** folder in the Downloads folder and drag onto the **Documents** location displayed in the sidebar/Navigation pane of the Finder window.

2. Click on the **Documents** location in the sidebar/Navigation pane to open the Documents folder.

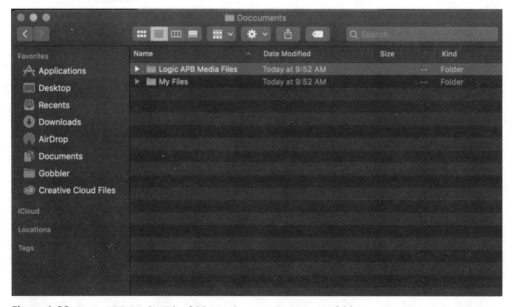

Figure 1.22 Logic APB Media Files folder in the open Documents folder

Listening to Sample Files

In this part of the exercise, you will locate and play various audio files. In the process, you will practice listening for different musical parts in each file.

Listen to the music mix:

1. Double-click on the **Logic APB Media Files** folder in the open Finder window to open the folder and display its contents.

2. Double-click on the **01. Pinder Brothers** folder to open it. You will see the following files:

   - OverboardMix.wav
   - Stem1.wav
   - Stem2.wav
   - Stem3.wav
   - Stem4.wav

3. Select the **OverboardMix.wav** file.

4. Press the spacebar to begin playback using the Mac's QuickLook feature.

5. As you play back the mix, listen for each of the different instruments and sounds that you can hear.

   Write down each part that you hear in the table below. Rate the prominence of the part in the mix on a scale from 1 to 5, with 5 being very prominent/easy to hear and 1 being barely noticeable.

Instrument or Part	Prominence
Lead Vocal	5

Exercise 1

Listen to the stem files:

1. Select the **Stem1.wav** file and listen to it using the same method that you used in Step 4 above.
2. Write down the main part you hear in the stem file, along with a short description, using the table on the next page.
3. Listen for additional parts in the stem file and indicate each different part you hear in the table along with a short description of each.
4. Repeat the process using each of the other stem files (Stem2, Stem3, and Stem4).

Not all of the stems will have different discernable parts. If you are not accustomed to picking out and describing the different musical sounds you hear, just do your best to make distinctions where something sounds different. This is good ear training regardless of what type of audio production you are interested in.

Some terms you might use in your descriptions include the following: lead vocal, background vocal, vocal harmony, vocal double, rhythm guitar, lead guitar, bass guitar, kick drum (bass drum), snare drum, crash cymbals, hi-hat, ride cymbal, etc.

Stem File	Description of the Instrument or Part
Stem1	Lead vocal – Main vocal melody throughout

Finishing Up

After you've listened to each of the stem files, return to the **OverboardMix.wav** file and listen through it again. Are you better able to distinguish each part in the mix, now that you've heard them in isolation?

You should now be able to clearly pick out things like the cymbal crashes and vocal harmonies that you may have missed before. You may be able to distinguish subtler details as well. Can you hear what the bass guitar is doing during the lead guitar part? Can you hear individual kick and snare drum hits? Can you tell where the hi-hat is being played and where it switches to the ride cymbal?

 To hear individual drum parts in isolation, check out the audio files in the z-Drum Parts folder. These files will help you learn what each drum piece sounds like and let you know what to listen for in a drum kit.

That completes this exercise. For more practice honing your listening skills, try identifying parts in popular songs that you hear on the radio or Spotify. If you are more interested in post-production applications, listen for different sound elements in your favorite TV program. Examples include dialog, background ambience (birds chirping, waves crashing), Foley (footsteps, character movements, noise from objects being handled), sound effects (telephones ringing, gun shots, laser beams), and background music.

Chapter 2

DAW Concepts

...*What You Need from Your Digital Audio Workstation*...

This chapter introduces you to the basic concepts of the DAW or Digital Audio Workstation. We begin by looking at some of the popular DAWs in use today. Then we explore the common audio plug-in formats that are used to add both effects and virtual instruments to a DAW. Next we consider the Logic Pro X system as a DAW for the Mac platform. We conclude this chapter with a brief explanation of some important terminology that you'll need to know to work effectively with Logic Pro X.

⊕ Learning Targets for This Chapter

- Learn the basic concepts of Digital Audio Workstations
- Gain an understanding of common plug-in formats
- Explore Logic Pro X
- Become familiar with important audio terminology

Key topics from this chapter are illustrated in the Logic Audio Production Basics Study Guide module available through the Elements|ED online learning platform. Sign up at ElementsED.com.

Although this book is written for Logic Pro X, it is important for you as an audio production enthusiast to have at least a passing familiarity with other digital audio products on the market. Not only are you likely to encounter audio practitioners who favor these platforms, you may end up using some of them yourself from time to time, to supplement your work in Logic Pro X.

Functions of a DAW

The term DAW stands for Digital Audio Workstation. This term is used generically to refer to any software that can be used to record, edit, and mix audio, although most modern systems can be used for MIDI recording and editing as well.

What Can a DAW Do?

Some DAWs excel at audio editing, while some are primarily known for their MIDI features, but almost every modern DAW can do both to some degree. Logic Pro X is generally considered to be well-balanced DAW that provides powerful tools for handling both audio and MIDI content. In addition, Logic Pro X offers a plethora of software instruments and plugins giving users everything they need to start producing right out of the box.

In addition to basic recording and editing functions for audio and MIDI, DAWs often provide sound modules (or virtual instruments) for use with MIDI data. Such virtual instruments are commonly available in the form of plug-ins that use MIDI data to trigger sounds.

Most DAWs also use plug-ins for audio processing, to apply effects and polish to audio recordings. Additionally, DAWs typically provide mixing and automation functions, for blending tracks, creating an effective stereo image, and incorporating dynamic changes during playback.

Common DAWs

Many DAWs are available, ranging in price and feature set. Here, we provide an overview of some of the most popular DAWs and highlight some of the differences between them.

Apple Logic Pro X

Logic Pro X is essentially the professional version of Apple's free entry-level DAW, GarageBand (discussed below). As an Apple product, it is available only for Mac OS computers, and it won't run on Windows computers. The product was originally developed by a company called Emagic, which got its start creating MIDI sequencing software for Atari computers.

Logic features a huge number of plug-ins effects including pro-level EQ, dynamics, and modulation effects. It also features a convolution reverb effect that can be used to create ultra-realistic models of real acoustic spaces.

The included virtual instruments in Logic are equally impressive. Logic includes a number of synthesizers, a physical modeling synth, a sampler, several drum instruments, as well as an organ and clavinet. Many MIDI composers and producers select Logic as their go-to DAW due to its robust support for virtual instruments.

Additional power-user features in Logic include Flex Time and Flex Pitch. These features allow users to freely adjust the timing and pitch of recorded audio.

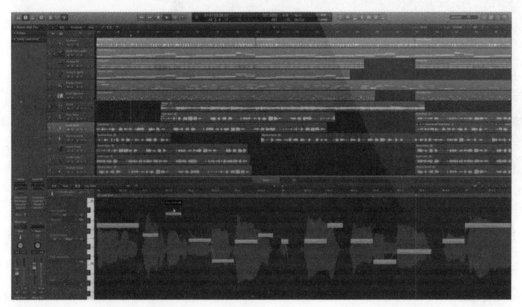

Figure 2.1 Flex Pitch editing in Apple Logic Pro X

Apple GarageBand

GarageBand is an entry-level DAW, designed to be easy to use for beginners. Like Logic Pro X, GarageBand runs only on Mac computers (no Windows version); however, GarageBand is also available on iOS platforms.

GarageBand makes it easy to get up and running to start making music right away. This DAW includes a nice complement of effects plug-ins including EQ, compression, reverb, and delay, as well as excellent guitar amp and pedal emulations.

The included virtual instrument plug-ins span the gamut from synthesizers to acoustic instruments to an automatic drum pattern generator. GarageBand also features a large selection of Apple Loops, which are professionally-recorded sound files that can be used to build a song very quickly.

The mobile version of GarageBand can run on both iPads and iPhones, making it easy for users to sketch out a song on-the-go, and then transfer the files to a Mac OS computer for further polishing. You can also import GarageBand songs directly into Logic Pro X.

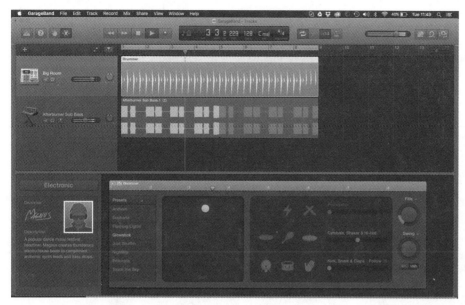

Figure 2.2 The Drummer instrument in Apple GarageBand

Avid Pro Tools and Pro Tools|First

Pro Tools became the first commercially successful DAW when it was introduced back in 1991. Since then, Pro Tools has evolved from a tool built exclusively for audio professionals into a popular DAW for beginners and professionals alike.

The Pro Tools audio editing feature set is largely unsurpassed in its power and simplicity, making it a standard in professional music production and audio post-production for film and TV. Pro Tools has also made great strides as a MIDI production tool, adding a number of advanced MIDI features over the years. Cross-integration with the Sibelius notation platform has made it easy to exchange MIDI information between Pro Tools and Sibelius, for sophisticated, professional-grade scoring.

 The Score Editor feature in Logic Pro X provides functionality similar to music notation in Pro Tools.

Pro Tools|First was released in 2015 as a free option for aspiring audio engineers. The product became a successful entry-level option for audio enthusiasts. Today, Pro Tools|First supports cloud collaboration workflows and interchange with standard Pro Tools software and Pro Tools|Ultimate systems.

Figure 2.3 A project in Pro Tools First

Ableton Live

Ableton's Live is a unique DAW that was created specifically as an electronic music performance tool. It offers a familiar track layout and audio and MIDI editing toolset in its Arrangement view. But where Live really stands out is the revolutionary Session view that took the DAW world by storm back in 2001.

In Session view, audio and MIDI clips are loaded into "slots" that can be triggered, stopped, and looped independently from one another. As a result, complex song arrangements can be performed on-the-fly in a way that isn't really possible with many other DAWs. This has made Live a favorite among live performers, DJs, and electronic musicians.

The inclusion of Cycling 74's MAX software with Ableton Suite (known as "Max for Live") offers another unique feature—the ability to create custom effects and virtual instrument plug-ins without writing any code.

In recent years, Live has probably seen more development of dedicated control surfaces than any other DAW, with Akai's APC series, Novation's Launchpad series, and Ableton's Push devices leading the way.

Figure 2.4 The Session View in Ableton Live

Steinberg Cubase/Nuendo

Steinberg's Cubase is one of the oldest and most respected DAWs on the market. The product began its development in the 1980s as a MIDI sequencer for the Atari ST (like Logic). Cubase has since evolved into a world-class DAW with outstanding MIDI capabilities and solid audio editing.

In particular, Cubase is very popular with hardcore MIDI users (such as film composers). This is due to several innovations that Steinberg introduced, including an ingenious chord track and elegant sampler articulation management. Additionally, the included VST plug-ins are top-notch.

 For information on VST and other plug-in formats, see "Plug-In Formats" later in this chapter.

Steinberg also manufactures another well-known DAW, by the name of Nuendo.

Nuendo is essentially an enhanced version of Cubase with more advanced audio features that target the audio post-production industry. Cubase and Nuendo both run on Mac OS and Windows computers, making either option an excellent choice for users who need to work on Windows or move between the two platforms.

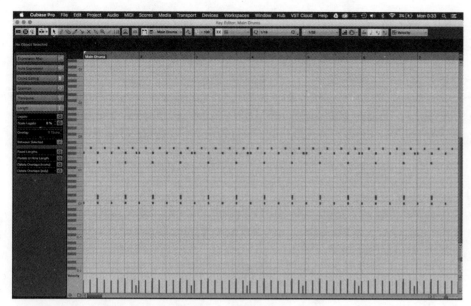

Figure 2.5 The powerful key editor in Steinberg Cubase

And the Rest...

Numerous other DAW options are available, with more coming to market everyday. Some examples of other popular DAWs include the following:

- **Cakewalk (Sonar)**—Cakewalk by Bandlab, formerly know as Sonar, evolved from the original Windows-only application that was a popular early MIDI sequencer. It now offers powerful audio functionality for both Windows and Mac computers, and is currently available for free.

- **Cockos REAPER**—REAPER is an inexpensive and flexible DAW. Many users are drawn to REAPER for its incredible customization options: REAPER allows users to create complex "actions" to accomplish almost any task. In addition, REAPER can be configured so that its editing features mimic the functionality of another DAW, making it easy to use for "switchers."

- **MOTU Digital Performer**—Digital Performer is another Mac-only DAW that has a loyal following in the film-scoring world. Its credentials include well-known users such as Danny Elfman. Digital Performer offers a number of editing tools that hold particular appeal for classically-trained composers and orchestrators, as well as a deep general toolset for MIDI and audio work.

- **PreSonus Studio One**—Studio One is a relatively young DAW that has garnered a lot of attention in recent years. It offers a nice balance of MIDI and audio tools and is constantly adding innovative new features.

- **Reason by Reason Studios (formerly Propellerhead)**—Reason was once an absolute necessity for any DAW user looking to add a cost-effective bundle of virtual instruments. Reason has since matured into a full-featured MIDI and audio editor. It continues to offer a huge range of virtual instruments with a user-friendly interface.

Plug-In Formats

Most DAWs support plug-ins for both audio processing and MIDI virtual instruments. But not all DAWs support plug-ins in the same format. As a result, a fair amount of confusion surrounds the issue of plug-ins and plug-in formats.

What Is a Plug-In?

A plug-in is a small software program that runs inside of your DAW to extend its functionality. Plug-ins are typically inserted onto a track in the DAW to run in real-time, although some plug-ins can also be used to directly process a selected portion of audio on a track.

Plug-ins fall into two general categories:

- **Effects Plug-ins**—Effects plug-ins offer a range of signal processing effects including equalization (EQ), dynamics processing (such as compressors and limiters), modulation (such as chorus and flange effects), harmonic processing (such as distortion), and time-based effects (such as reverb and delay).

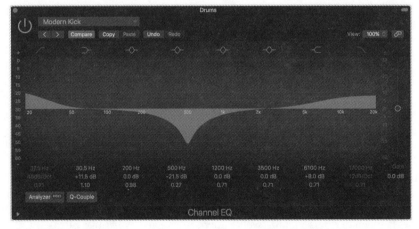

Figure 2.6 Logic's Channel EQ is an example of an effects plug-in

- **Virtual Instrument Plug-ins**—Virtual instrument plug-ins are software emulations of hardware musical instruments. These instruments include synthesizers (both analog emulations and modern digital algorithms), samplers, and drum machines.

Figure 2.7 Alchemy is an example of a virtual instrument plug-in.

Common Plug-In Formats

Some common plug-in formats include the AU format developed by Apple and the VST format developed by Steinberg. The AAX format, developed by Avid, is used exclusively for Pro Tools systems and other Avid products. Although discontinued, the RTAS and DSP formats used by older Pro Tools systems are also still common in many Pro Tools-based production facilities.

The following breakdown provides details on each of these formats:

- **AU**—The Audio Units (AU) format is Apple's audio plug-in format. AU plug-ins are used primarily in Logic Pro X and GarageBand, but they are also supported in several other DAWs (including Ableton Live) as well as a number of simpler audio applications on Mac OS.

- **VST**—Virtual Studio Technology (VST) is a plug-in format created by Steinberg. It was originally developed for Cubase in the 1990s, but the format was rapidly adopted in most of the other DAWs of that era. VST is easily the most popular format for third-party plug-ins and is supported in every major DAW except for Logic Pro X, GarageBand, and Pro Tools.

- **AAX**—The Avid Audio eXtensions (AAX) specification is the current plug-in format supported in Pro Tools. This specification supports both AAX Native and AAX DSP formats. AAX is the only plug-in format supported in Pro Tools 11 and later. AAX plug-ins are not currently supported in any other (non-Avid) DAW.

- **AAX AudioSuite**—The AAX AudioSuite format is Avid's plug-in format for rendered plug-in processing. When AudioSuite processing is applied, a new file is rendered with the effect written in. AudioSuite processing cannot be changed in real-time like other plug-in processing.

- **RTAS and TDM**—Real-Time AudioSuite (RTAS) and Time-Division Multiplexing (TDM) are obsolete plug-in formats that were created by Avid (then Digidesign). RTAS plug-ins are host-based processors that run on the computer's CPU, while TDM plug-ins are DSP-based and run on custom Motorola DSP chips included on Pro Tools DSP cards. These plug-in formats are supported through Pro Tools 10; support was discontinued beginning with Pro Tools 11.

Logic Pro X System

The latest update of Logic Pro X is version 10.4.8 (as of this writing). Notable capabilities and features are listed below. A comprehensive list can be found at Apple's website.

Logic Pro X Software Capabilities

Version 10.4.8 of Logic Pro X provides the following capabilities:

- Up to 1000 audio channel strips (mono or stereo)
- Up to 1000 auxiliary channel strips (mono or stereo)
- Up to 1000 software instrument channel strips (mono or stereo)
- Up to 1000 external MIDI tracks
- Up to 256 busses
- 15 inserts for internal or Audio Unit effects plug-ins
- 8 inserts for internal or Audio Unit MIDI plug-ins
- 12 sends per channel strip, pre- or post-fader, or post-pan
- Audio file and I/O resolution up to 24-bit/192kHz
- Non-destructive, random-access editing and mix automation
- Ability to open GarageBand songs directly
- Support for surround sound configurations
- ReWire support for Reason, Ableton Live, and other compatible applications
- 1220 definable key and MIDI commands

Logic Pro X Software Options

Version 10.4.8 of Logic Pro X includes the following additional software options:

- 69 effect plug-ins including Pedalboard, with 35 stompboxes
- 23 software instrument plug-ins
- 9 MIDI plug-ins
- 2900 Patches for Audio, Auxiliary, Software Instrument, and Output tracks
- 10,000 royalty-free Apple Loops covering a wide range of genres
- 1000 EXS24 Sampler instruments
- 660 reverb spaces and warped effects for Space Designer

Logic Pro X Hardware Options

Logic Pro X works with Core Audio–compliant audio devices, including Thunderbolt, FireWire, USB, ExpressCard, and PCI audio interfaces. Logic Pro X automatically recognizes any installed Core Audio hardware, making it easy to plug-and-play with most audio interfaces.

Figure 2.8 Focusrite's Scarlett 2i2 is an excellent entry-level interface.

Figure 2.9 MOTU's UltraLite Mk3 is an affordable, mid-range interface.

Downloading and Installing Logic Pro X

If you are ready to proceed with Logic Pro X as your DAW of choice, you will need to start by installing the software. Several steps are required to download and install Logic Pro X.

System Requirements for Logic Pro X

Before installing Logic Pro X, confirm that your system meets the following requirements:

- macOS 10.13.6 or later
- 4GB of RAM
- OpenCL-capable graphics card or Intel HD Graphics 3000 or later
- 256MB of VRAM
- 6GB of available disk space for a minimum installation, or up to 63GB of disk space for the full Sound Library installation

If you are running an operating system that is earlier than 10.13.6, you will not be able to install the latest version of Logic Pro X. You may, however, get a message box asking if you would like to install the last version compatible with your system. In that case, the installation instructions will be similar but may vary from the steps shown below.

Installation Steps

To install Logic Pro X, open the Mac App Store and follow the steps below:

1. **Create an Apple ID**—An Apple ID is required to download Apple products. In this first step, you will create an account for free on the Apple website. (See Figure 2.10.) If you already have an Apple ID, proceed to step 2.

Figure 2.10 Step 1: Create an Apple ID

2. **Log in to the Mac App Store with your Apple ID**—Open the Mac App Store application and log in with the credentials for your account.

Figure 2.11 Step 2: Log in to the Mac App Store

3. **Buy or Get Logic Pro X**—Browse the App Store or use the search bar to find Logic Pro X. Place your mouse over the blue price to reveal the **BUY** button to make your purchase. (See Figure 2.12.) If you have already purchased the software, this button will be replaced by a **GET** button or cloud-download icon.

Figure 2.12 Step 3: Purchase Logic Pro X from the App Store

4. **Download and install Logic Pro X**—Finally! In this last step, you will download Logic Pro X. The blue button will change to a **GET** button cloud-download icon. After clicking the button, Logic Pro X will begin downloading and will automatically install to your Mac.

   Once this process is complete, you will be able to launch Logic Pro X from the App Store or from your Applications folder. (See Figure 2.13.)

Figure 2.13 Step 4: Open Logic Pro X from the App Store

Launching Logic Pro X for the First Time

The first time you launch Logic Pro X, you will be prompted with a message box with links to additional information for beginners and descriptions of features of the current version. (See Figure 2.14.) A collection of essential audio resources will also begin downloading. This download provides a selection of sounds, instruments, and loops that let you start making music immediately.

You can click the **DOWNLOAD LATER** button, if desired, to cancel the download and complete it at a later time, or wait for the download to complete.

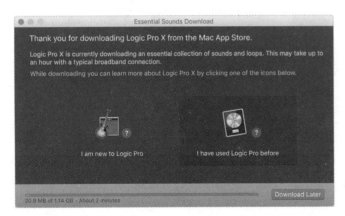

Figure 2.14 Essential Sounds Download Message

Once you've proceeded past the initial dialog box, you will be presented with the Project Chooser. This dialog box lets you create a new project or open an existing one. For now if you'd like to explore, you can choose to create a new project. Details on working with projects are provided starting in Chapter 5.

Downloadable Content

Logic Pro X comes with a large selection of Apple Loops, patches, drum kits, plug-in settings, sampler instruments and their settings, impulse responses, and legacy content that you can use in your projects. Essential content begins downloading on first launch of the software. You can select additional content to be downloaded in the Sound Library Manager by choosing **LOGIC PRO X > SOUND LIBRARY > OPEN SOUND LIBRARY MANAGER**. (See Figure 2.15.)

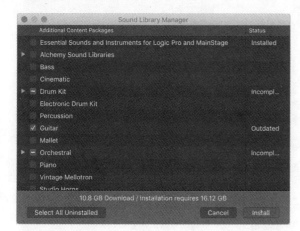

Figure 2.15 Select additional content to download in the Sound Library Manager

 If you canceled the content download during the initial launch, you can restart it in the Sound Library Manager by selecting Essential Sounds and Instruments.

Important Preference Settings

While working on a project, you may need to adjust certain preferences and other settings. The Preferences dialog box can be accessed by selecting **LOGIC PRO X > PREFERENCES**. Along the top of the dialog box are various categories to choose from, including General, Audio, Recording, MIDI, and more. To get started, you'll need to configure your audio device settings. You may also want to enable Advanced Tools to access all the tools Logic Pro X has to offer.

- **Audio Settings:** In the Devices tab of the Audio Preferences, you can select individual input and output devices for Logic. (See Figure 2.16.) These selectors let you choose between the built-in audio capabilities of your computer and a connected audio interface. When you select a new device, you'll need to click the **APPLY CHANGES** button to load the audio drivers for the updated settings.

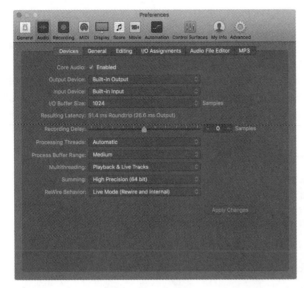

Figure 2.16 Select an audio output device and input device in Audio Preferences.

- **Recording Bit Depth:** By default, Logic Pro X records audio with 24-bit resolution. You have the option to change to 16-bit by deselecting **24-Bit Recording** in the Recording Preferences.

Figure 2.17 Adjust audio bit depth for recording

- **Advanced Tools:** By default, Logic Pro X limits the amount of tools and features that are visible from the initial launch of the software. (The look is reminiscent of GarageBand.) To get the most out of your Logic Pro X experience, it is recommended that you enable the **Show Advanced Tools** option in Advanced Preferences and click the **Enable All** button to access all available features. (See Figure 2.18.)

 For this course, you will need to enable the Advanced Tools (Logic Pro X > Preferences > Advanced Tools). Be sure to check all of the available boxes, so that everything is enabled, as shown in Figure 2.18.

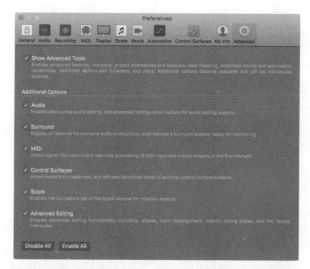

Figure 2.18 Show Advanced Tools enabled in Preferences

Project Settings

While Preference settings apply to the current project and all future projects (until changed), changes you make to Project Settings apply only to the active project. The Project Settings dialog box can be accessed by selecting **File > Project Settings**.

- **Sample Rate:** By default, Logic Pro X uses a sample rate of 44.1 kHz. You can change this setting for your project, selecting from common sample rates up to 192 kHz, under the Audio tab in Project Settings. (See Figure 2.19.)

Figure 2.19 Sample Rate selector in the General Audio pane of Project Settings

 Logic Pro X supports importing audio files with any standard bit depth and sample rate without the need for conversion.

Important Concepts in Logic Pro X

You should master a few key DAW concepts before getting started with audio software such as Logic Pro X. These include understanding how the software handles audio files for your projects and recognizing the fundamental differences between audio and MIDI data.

Project Files Versus Audio Files

Logic Pro X manages two main file types: project files and audio files. It is important to recognize the relationship between these file types.

- **Project Files**—The first thing you'll do when you start working with Logic Pro X is to create a new project or open an existing project. A Logic Pro X project file contains all of the information about the current audio production. This information includes the number and type of tracks in the project, the position of audio and MIDI data on each track, the plug-in assignments used on each track, the mixer level settings, the project settings, and a whole lot more.

    You will typically create a unique project file for each song or production you work on.

    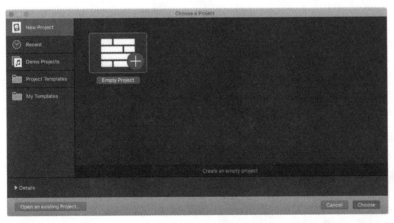

    Figure 2.20 The Project Chooser dialog box for creating a new project or opening an existing project

- **Audio Files**—An audio file represents a single audio recording (or *take*) that you've either created directly in your project or imported into the project from a disk location. A typical project will have dozens or even hundreds of audio files associated with it.

    Logic Pro X stores audio files as WAV files by default, but many other types of audio files can be imported into your project including AIFF, CAF, MP3, QuickTime, Audio CD, and more.

> ### Referencing Versus Copying Imported Audio Files
>
> Logic Pro X gives you the option to copy audio files into your project or reference them in another location. While referencing audio files reduces redundancies and saves space, copying audio files ensures your project will run smoothly if moved to another location. When saving your project, you can select whether or not audio files will be copied. You can change this setting later by choosing FILE > PROJECT SETTINGS > ASSETS.

Audio Versus MIDI

The differences between audio and MIDI data are very important to understand when you're just starting out in the world of audio production. Audio files represent the audio signal that was recorded when the sound (from a voice, instrument, or other source) was captured using an audio interface. Therefore, the audio file itself actually *contains* the sound information that was captured. Audio files can be played back through the audio interface, allowing the sound to be heard using monitor speakers or headphones.

On the other hand, MIDI data is a series of note and control messages, typically recorded from a MIDI controller (like a keyboard or drum pads). MIDI data can also be manually entered in Logic Pro X. The resulting MIDI file does NOT actually contain any sound information. The MIDI file's note and control messages must be sent to a real or virtual instrument, which can then turn the messages into an audio signal.

Figure 2.21 A Software Instrument track with a MIDI region (top) and an Audio track with two audio regions (bottom)

Review/Discussion Questions

1. What are some tasks that can be completed using a Digital Audio Workstation (DAW)? (See "Functions of a DAW" beginning on page 32.)

2. What are some of the differences between the common DAWs? How are they similar? (See "Common DAWs" beginning on page 32.)

3. What is an audio plug-in? How are plug-ins enabled in a DAW? (See "What Is a Plug-In?" beginning on page 38.)

4. What are the two general categories of plug-ins? Give some examples of plug-ins in each category. (See "What Is a Plug-In?" beginning on page 38.)

5. Which of the common plug-in formats is/are supported in Logic Pro X? (See "Common Plug-In Formats" beginning on page 39.)

6. What are some audio interfaces that are supported in Logic Pro X? (See "Logic Pro X Hardware Options" beginning on page 41.)

7. What are some options you'll be presented with when first launching Logic Pro X? (See "Launching Logic Pro X for the First Time" beginning on page 44.)

8. What are the two primary types of files that Logic Pro X manages? (See "Important Concepts in Logic Pro X" beginning on page 48.)

9. What are some of the differences between audio and MIDI information? (See "Audio Versus MIDI" beginning on page 49.)

 To review additional material from this chapter and prepare for certification, see the Logic Audio Production Basics Study Guide module available through the Elements|ED online learning platform at ElementsED.com.

Exercise 2

Setting Up a Multi-Track Project

🎧 Activity

In this exercise, you will create a simple project containing the main parts for the song "Overboard" by Sacramento-area band, The Pinder Brothers. You will then configure the project settings for playback and play through the project.

🕓 Duration

This exercise should take approximately 10 minutes to complete.

⊕ Goals/Targets

- Launch Logic Pro X
- Use the Project Chooser to create a blank project document
- Import audio to the project
- Use the Audio Device Preferences to configure the project for playback
- Save your work

Exercise Media

This exercise uses media files from the song, "Overboard," provided courtesy of The Pinder Brothers.

Written by: Matt Pinder; Performed by: The Pinder Brothers;
*Produced by: Scott Reams and The Pinder Brothers**

The media provided for this course may be used for educational purposes only. No rights are granted to use the media for any other personal, commercial, or non-commercial purposes.

* *The mix, processing, and media files have been adapted for use in the exercises contained herein.*

52 Exercise 2

Getting Started

To get started, you will launch Logic Pro X as described in Chapter 1. Then you will use the Project Chooser to create a new project.

 Before starting this exercise, Logic Pro X must be installed on your system. If necessary, complete the Logic Pro X download and installation steps, as outlined in Chapter 2.

 If you haven't done so already, be sure to enable all Advanced Tools before proceeding (LOGIC PRO X > PREFERENCES > ADVANCED TOOLS).

Launch Logic Pro X and create a project:

1. Launch Logic Pro X by clicking on the Logic Pro X icon in the Dock (or using another method of your choice).

2. When the Logic Pro X Project Chooser appears, select **EMPTY PROJECT**.

    If the Project Chooser isn't visible, choose FILE > NEW FROM TEMPLATE. This will open the Project Chooser. Then select Empty Project and proceed.

3. Click the **CHOOSE** button at the bottom of the Project Chooser to create the project. A new project will be created and will display on screen.

4. Logic Pro X will prompt you to create a new track before continuing. For this exercise, create one new Audio track.

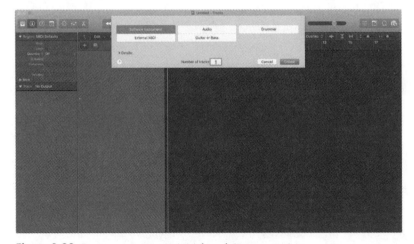

Figure 2.22 Prompt to create an initial track in your project

> Logic requires that there always be at least one track in your project. If you decide not to use the track you initially create, you can delete it after adding other active tracks.

Importing Audio for the Project

In this section of the exercise, you will configure the Main window and import audio to the project. The process you'll use will automatically create tracks for the imported audio files.

Configure the Main window:

1. Choose **WINDOW > OPEN MAIN WINDOW**, if needed, to bring the Main window to the forefront and make it active.
2. Resize or maximize the Main window for a full-screen view.

Import the project audio:

1. Open the Browser by clicking the **BROWSER** button (media icon) on the far right of the Control bar (or press the **F** key).

Figure 2.23 The Browser button in the Control bar

2. Select the **ALL FILES** tab in the Browser, and navigate to your Documents folder (or other location where you saved the **Logic APB Media Files** folder in Exercise 1). Optionally, you can navigate to this folder using the Finder.
3. Open the 02. **Project Files** folder within the **Logic APB Media Files** folder.
4. Select the thirteen audio files in the folder. (See Figure 2.24.) Drag the selected files onto the first measure in the Tracks area of the Main window in Logic.

 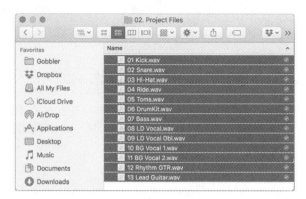

Figure 2.24 Files selected in the Browser (left) and in a Finder window (right)

5. When the **ADD SELECTED FILES TO TRACKS** dialog box displays, select the **USE EXISTING TRACKS** radio button.

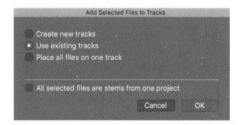

Figure 2.25 Add Selected Files to Tracks dialog box

> If you select **CREATE NEW TRACKS**, you will leave your initial audio track empty.

6. Click **OK** to complete the import. Twelve additional tracks will be created in your project.

    The imported audio will be placed on your thirteen tracks, starting with the initial audio track you created previously. (See Figure 2.26.)

Setting Up a Multi-Track Project   55

Figure 2.26 Main window with imported tracks

Playing the Project

In this part of the exercise you will configure the Audio Devices Preferences and play back the project.

Play back the project:

1. Choose **LOGIC PRO X > PREFERENCES > AUDIO** to display the Audio Devices Preferences.

2. Verify that your connected audio interface is displayed in the Output Device pop-up menu (if applicable). If needed, click on the pop-up menu to select the interface.

Figure 2.27 Built-in Output is selected as the Output Device

> ⓘ Changing the selected audio interface requires that you click Apply Changes to reload the audio drivers and commit the changes. This is normal behavior.

3. Press the **SPACEBAR** to begin playback. The Playhead cursor will scroll across the screen and you will hear playback (if you have speakers or headphones connected to your system).

4. While playing back the project, scroll the Main window vertically, as needed, to view the audio waveforms on each track. You can use the scroll bar on the right side of the window or another method of your choice to scroll the window.

5. When finished, press the **SPACEBAR** a second time to stop playback.

Finishing Up

To complete this exercise, you will need to save your work and close the project.

Finish your work:

1. Choose **FILE > SAVE AS** to save the project.

2. Name your project Overboard01-*xxx*, where *xxx* is your initials, and select an appropriate save location. Under **ORGANIZE MY PROJECT AS A:** choose the **FOLDER** radio button and press **SAVE**.

3. Choose **FILE > CLOSE PROJECT** to close the project. (Or click the red button on the top left of the Main window.)

> ⓘ You can close a project by closing the Main window as long as there are no other active windows associated with your project.

4. If desired, you can choose **LOGIC PRO X > QUIT LOGIC PRO X** to exit from the software.

That completes this exercise.

Chapter 3

Audio Recording Concepts

...*What You Need to Record Audio*...

This chapter introduces you to the fundamentals of audio recording. We begin with a discussion on the basics of sound, including the principles of frequency and amplitude. We then discuss various types of microphones and their uses, before diving into a discussion on multi-track recording. We follow this by examining the process of converting audio signals between the analog and digital domains and exploring the role of the audio interface in this exchange. The characteristics of analog and digital audio discussed in this chapter are important considerations when it comes to optimizing your results with any DAW.

Learning Targets for This Chapter

- Understand the audio characteristics of frequency and amplitude
- Understand characteristics of traditional microphones and USB microphones
- Understand the purpose of multi-track recording
- Understand the basic principles of analog-to-digital conversion
- Recognize the role of the audio interface in modern digital recording

Key topics from this chapter are illustrated in the Logic Audio Production Basics Study Guide module available through the Elements|ED online learning platform. Sign up at ElementsED.com.

The process of recording audio has been commonplace for over 60 years, ever since magnetic tape-based recording processes began replacing earlier processes of inscribing a signal to a physical surface. In just the past 20 years or so, tape-based recording has itself largely given way to digital recording technology.

Regardless of the technology used to capture an audio recording, certain fundamental characteristics remain unchanged, including how an audio event moves through an acoustic environment and how that event is translated into an electrical signal that can be recorded. Recording in digital simply adds a dimension of collecting discrete measurements of the signal and storing those measurements as binary information.

The Basics of Audio

To understand the process of recording audio, it helps to first understand the basics of audio and sound waves. Sound waves are created when a physical object vibrates, causing a variation in the surrounding air pressure. By way of example, consider what happens when a guitar string is plucked.

The vibration of the guitar string, as it moves back and forth in a repeating cycle, causes a displacement of air particles. This results in a cyclical variation in air pressure, known as compression and rarefaction. The air pressure will increase (compression) and decrease (rarefaction) in a way that corresponds directly to the pattern and frequency of the string's vibration. This cycling pattern of air pressure changes is referred to as a sound wave.

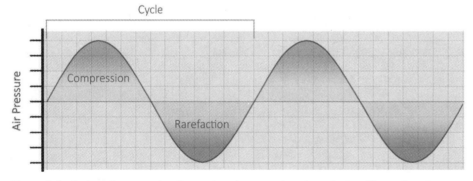

Figure 3.1 Cycles of compression (increasing air pressure) and rarefaction (decreasing air pressure) in a sound wave

Frequency

The frequency of the cycle of sound pressure variations determines our perception of the pitch of the sound. If a guitar string vibrates quickly, the air pressure variations will likewise cycle quickly between compression and rarefaction, creating a high-pitched sound. If the guitar string vibrates slowly, the air pressure will cycle slowly, creating a low-pitched sound.

We measure the frequency of these changes in *cycles per second* (CPS), also commonly denoted as *Hertz* (Hz). These two terms are synonymous—15,000 CPS is the same as 15,000 Hz. Multiples of 1,000 Hz are often denoted as kilohertz (kHz). Therefore, 15,000 Hz is also written as 15 kHz.

The open "A" string on a guitar vibrates at a frequency of 110 Hz in standard tuning. Playing the A note on the 12th fret produces vibrations at 220 Hz (one octave higher).

The range of human hearing is generally accepted to be between 20 and 20,000 cycles per second. This is also commonly denoted as 20 Hz to 20 kHz. Therefore, to capture a full-spectrum recording, we must be able to represent all frequencies within this range.

Amplitude

The intensity of a sound, or the amount of change in air pressure it produces, creates our perception of the loudness of the sound. A sound's intensity is represented by the amplitude (height) of the sound wave. We measure amplitude in *decibels* (dB).

The decibel scale is defined by the dynamic range of human hearing, from 0 dB at the threshold of hearing to approximately 120 dB at the threshold of pain. The decibel is a logarithmic unit that is used to describe a ratio of sound pressure. As such, the decibel does not have a linear relation to our perception of loudness. An increase of approximately 10 dB is required to produce a perceived doubling of loudness.

By way of example, the amplitude of ordinary conversation is around 60 dB. Increasing the amplitude to around 70 dB would essentially double the loudness, similar to what you might face trying to hold a conversation in a room with the TV or radio on. Increasing the amplitude to 80 dB would double the loudness again, such as you might experience trying to hold a conversation in a crowded room.

Microphones

The most common technique used to capture an acoustic audio event for recording is to place a microphone somewhere near the sound source. Microphones come in many shapes, sizes, and prices, so it can help to know a little bit about how they work before selecting the microphone or microphones you plan to use.

Traditional Microphones

All microphones are transducers, meaning they change energy from one form to another. A traditional analog microphone simply converts the variations in air pressure into variations in an electrical current. Three basic types of microphones are commonly available to make this conversion: dynamic mikes, ribbon mikes, and condenser mikes.

Dynamic Microphones

A dynamic microphone consists of a diaphragm attached to a coil of wire encircling a magnet. When sound waves hit the diaphragm, it vibrates back and forth, thereby moving the coil of wire back and forth over the magnet. This movement induces a current in the coil, with compression creating a positive voltage and rarefaction creating a negative voltage.

The advantages of dynamic microphones are that they are rugged and durable. They can also handle high input signals without distortion. Dynamic microphones are commonly used in both studio and live performance environments.

Ribbon Microphones

A ribbon microphone consists of a thin metal foil or other conductive ribbon material suspended between the positive and negative poles of a magnet. As sound waves hit the ribbon, they cause it to vibrate; the ribbon's movement within the magnetic field induces a current within the ribbon itself.

The advantages of ribbon microphones include a warm, smooth tone and an ability to capture high-frequency detail. Some ribbon microphones can be delicate and expensive. Ribbon mikes tend to be better suited for use in a studio environment rather than live performance.

Condenser Microphones

Condenser microphones consist of a thin conductive diaphragm (front plate) affixed a small distance in front of a metal back plate. The two plates are typically energized with a fixed charge. When sound waves hit the front plate, the diaphragm vibrates, varying the distance between the two plates. This varies the capacitance of the circuit and creates an opposite change in electrical voltage.

The capacitance is the ratio of the electric charge on the plates to the voltage difference between them. The capacitance increases as the plates move closer together and decreases as the plates move further apart.

Condenser microphones require power to operate, either from a battery or a phantom power source. Phantom power is typically 48 volts DC applied through a microphone cable to the condenser mike.

Mixers, microphone preamps, and audio interfaces commonly provide the option to enable phantom power, often denoted as +48V on the device.

The advantages of condenser microphones include a wide, smooth frequency response and sharp, detailed transient response. Condenser mikes are particularly well suited for acoustic instruments and cymbals. Although condenser mikes are very popular for in-studio recording, they are also commonly used in live sound production as well, particularly as drum overhead mikes.

USB Microphones

Another option to consider for home or studio use with your DAW is the USB microphone. USB mikes function the same as their traditional microphone counterparts, in that they convert acoustical energy into electrical energy. However, a USB mike goes a step further using two additional circuits: a built-in preamp and an analog–to–digital converter. This additional circuitry allows the microphone to be plugged directly into your computer for use with your DAW without requiring a separate audio interface.

For information on the analog-to-digital conversion process, see "Moving Audio from Analog to Digital" later in this chapter.

Most USB microphones are condenser mikes, although a few dynamic USB mikes are also available.

Popular choices among USB microphones include:

- **Blue Microphones**
  - **Snowflake USB Microphone**—Designed for mobile desktop recording at an affordable price ($)
  - **Snowball USB Microphone**—Designed for podcasting and other desktop recording ($$)
  - **Yeti USB Microphone**—Designed for studio vocals, musical instruments, voiceovers, field recordings, podcasting, and desktop recording ($$$)
  - **Spark Digital Lightning Microphone**—Designed for studio and desktop recording for voiceovers, vocals, and instruments at a premium price ($$$$)
- **CAD U37 USB Studio Condenser Recording Microphone**—Designed for desktop recording and voiceovers, vocals, and instrumental recording on a budget ($)
- **Audio-Technica ATR2100-USB Cardioid Dynamic USB/XLR Microphone**—Designed for stage and studio use at a reasonable price ($$)
- **Rode NT-USB USB Condenser Microphone**—Designed for studio-quality recording of vocals, instruments, and voiceovers with built-in monitoring control ($$$)
- **Apogee Mic 96k Professional Quality Microphone**—Designed for high-resolution studio-quality recording at a premium price ($$$$)
- **Rode Podcaster USB Dynamic Microphone**—Designed for studio-quality podcasting and other desktop recording at a premium price ($$$$)

Other Considerations

Aside from the characteristics of different types of microphones described above, a few other considerations should come into play when selecting the right mike for your needs.

Polar Pattern

A microphone's polar pattern describes how the microphone responds to sound coming from different directions. Typical choices include omnidirectional mikes, bidirectional mikes, and unidirectional mikes.

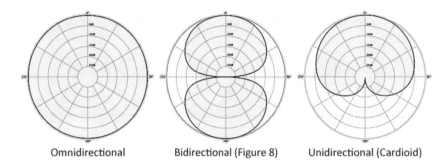

Figure 3.2 Common polar patterns: omnidirectional, bidirectional, and unidirectional

Omnidirectional. An omnidirectional mike will pick up sound equally from all directions. While this may be desirable for a microphone placed in the middle of a large conference table, it is usually not the best choice for studio recording where the goal is to isolate a sound source.

Bidirectional. Bidirectional mikes pick up sound primarily from the front and rear of the microphone, rejecting sound coming from the sides. This kind of mike may be desirable for across-the-desk interviews, two-part vocal recordings, or similar recordings of two opposite-facing sound sources. By rejecting sound from the sides, this type of mike will limit the amount of room acoustics and unwanted ambient noise in the recording.

Unidirectional (cardioid). Unidirectional mikes are the most popular choice for recording isolated sound sources. These are also commonly referred to as cardioid microphones. Unidirectional, or cardioid, polar patterns are designed to pick up sound from the front of the microphone, while rejecting (or attenuating) sound coming from the sides or rear of the microphone.

Proximity Effect

Most unidirectional and bidirectional microphones exhibit a noticeable bass boost when used within a few inches of a target sound source. This phenomenon is known as the proximity effect and often results in an unwanted boomy sound. You can hear the effect by speaking into a microphone as you move up close to the front of the grill.

If you are aware of a mike's proximity effect, you can adjust the mike placement or roll off the bass frequencies with an equalizer to compensate.

Basic Miking Techniques

Microphone placement and position have a large impact on how well a microphone will perform at capturing the desired signal. The overall goal when setting up a microphone is to target the sounds you want to record while rejecting the sounds you don't. Several accessories are available to help with this process.

Microphone Accessories

The microphone accessories discussed below can assist you with mike placement, noise reduction, and sound isolation.

- **Mike Stand**—USB microphones designed for desktop use often come with a small stand or clip. For miking in a studio or open environment, however, you will need to invest in a full-sized mike stand. You might also consider a stand with a boom extension or gooseneck add-on for better reach and flexibility. You'll find a variety of stands and extensions readily available from your local music store or online retailer.

- **Windscreens and Pop Filters**—Windscreens and pop filters are designed to reduce unwanted noise resulting from wind or bursts of air hitting the microphone. These problems are especially prevalent for on-location field recordings and close-miked recordings of vocal or dialog performances.

> In vocal performances, bursts of air are commonly produced from words that begin with "p" or "b" sounds. These bursts, or plosives, can ruin a recording if they hit a microphone's diaphragm directly.

- **Shock Mounts**—Shock mounts are designed to suspend a microphone, isolating it from any mechanical vibrations that may affect the mike stand. Such vibrations are commonly caused by loud music, noise carried through a floor or stage, or nearby thumps and bumps.

- **Acoustic Shields and Vocal Booths**—To improve sound isolation, especially for vocal and dialog recordings, you can consider using an acoustic shield. These products are designed to reduce the influence of ambient noise, room acoustics, and reflected sound on a recording. Acoustic shields are most effective at attenuating mid– and high–frequency audio.

  A more extreme approach is to build or install a fully–enclosed vocal booth or isolation room. These will be more effective at reducing unwanted noise than a shield, but they can also be expensive and complex to design.

  Note, also, that both shields and booths may colorize the audio, boosting or cutting certain frequencies of the performance. You may be able to correct for the colorization using an equalizer; however, you will need to determine whether the isolation gains are worth the trade-off.

Where to Shop for Audio Gear

While locally owned music stores are becoming scarce, audio gear has never been easier to shop for. In addition to ubiquitous Guitar Center storefronts, numerous online retail outlets are available. Online options include guitarcenter.com, sweetwater.com, musiciansfriend.com, vintageking.com, and others. Almost any audio equipment you need can be found with a quick Internet search!

Basic Setup and Connections

Setting up your microphone or microphones is a simple matter of finding the right mike position and establishing a connection to your computer.

- **Microphone Position**—The location of your microphone during a recording can have a dramatic influence on the recorded signal. Generally speaking, the closer the mike is to the target sound source, the "cleaner" the recording will be. That is to say, the recording will include more of the desired sound and less ambient noise.

   However, as noted above, certain mikes exhibit a proximity effect when placed close to their sound source. This can colorize the sound in ways that may be unwanted, depending on the situation.

- **Microphone Connection**—The way that your microphone connects to your computer will depend on the type of mike you are using. Most analog microphones use a cable with XLR connectors (3-pin) to attach to your audio interface. The audio interface will in turn connect to your computer using a USB cable or other common digital computer connection.

Figure 3.3 Microphone XLR connection to an Mbox Pro audio interface

A digital USB microphone will connect directly to your computer, with no audio interface required.

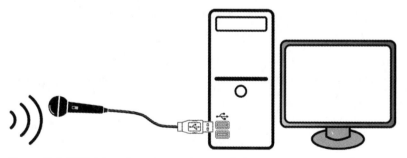

Figure 3.4 USB microphone connection to a desktop computer

Multi-Tracking and Signal Flow

The benefits of recording to a DAW are many. For starters, it allows you to record multiple takes and keep the best parts of each take. Additionally, it gives you the ability to edit and process a recording to improve the final results. But one of the greatest advantages is the freedom to record numerous isolated parts, each on its own track (known as multi-track recording).

What Is Multi-Track Recording?

The concept of multi-track recording has been around for many decades and is a fundamental aspect of audio production. Using this production method, each audio source is isolated as an individual recording on its own track. Each of these recorded tracks can then be edited and processed discretely. By playing back all of the individual tracks simultaneously, you can create a mix of the multiple sound sources as a final output.

By way of example, a music recording might include individual tracks for vocals, guitar, bass guitar, keyboards, and drums. By isolating each part on its own track, the volume levels, pan settings, EQ settings, and other processing can be set independently for each component.

Another advantage of multi-track recording is that individual parts can be recorded at different times. This is key if the performers cannot all be present in the same location at the same time. It also allows individual contributors to record multiple parts. For example, a musician can record both guitar and piano parts for a song. Or a voice actor can record dialog lines for multiple characters in a script for an animated short.

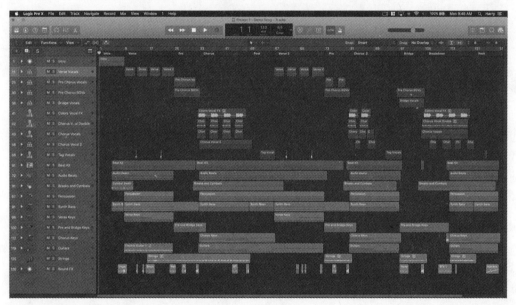

Figure 3.5 Multi-track music production session in Logic Pro X

Not all recordings require multiple tracks. In some cases, you may want to record a mix of multiple sources to a single track. Examples might include recording a live ensemble performance with a pair of microphones routed to a single stereo track or using an external mixer to sum multiple drum microphones to a single track. One reason for doing this is to simplify the recording and the number of tracks in your session. You might also use this approach if you do not have enough inputs on your system to record from many microphones at the same time.

Recording Signal Flow

Recording onto a computer involves routing a signal through various processing stages. This routing path is commonly referred to as the signal flow. Here, we will focus on a typical recording signal flow, from the sound source to the storage device (HDD or SSD).

The sound originates at a source, such as a guitar, and travels through an environment by way of variations in air pressure, as discussed above. The air pressure changes are picked up by a microphone, converting the sound into an electrical signal. The electrical signal travels down the microphone cable to an audio interface.

Within the audio interface, the signal is boosted, using a pre-amplifier (or *pre-amp*). The pre-amp is used to create a healthy signal level for recording. Next, the signal is sampled, using periodic measurements of the electrical voltage. The measured values are represented as strings of binary numbers and are passed along to the computer. At the computer, the binary numbers are stored on a hard disk drive or solid-state drive.

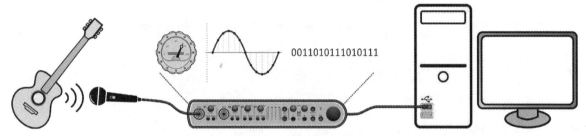

Figure 3.6 Recording signal flow from a guitar to a computer's hard drive

Moving Audio from Analog to Digital

As mentioned above, the practice of recording audio to a computer involves converting an electrical signal into individual, discrete binary measurements. This is the process known as analog-to-digital conversion.

Analog Versus Digital Audio

The variations in electrical voltage produced by a microphone capture a continuously changing analog audio waveform. In order to represent that information on a computer, the waveform must be converted into binary numbers that the computer can understand.

> ### What Is Binary Data?
> Binary data is comprised of binary digits, or bits. Each binary digit can represent only two possible values: *zero* or *one*. (At the most basic level, computers store and work with values using "switches" that are either off or on. A bit value of *zero* can be stored by toggling the associated switch off, while a bit value of *one* can be stored by toggling the switch on.)

> In order to store values larger than one, computers use strings of multiple binary digits. For example, using a string of four binary digits, a computer can store values from zero (0000) to fifteen (1111). Using a string of eight binary digits, a computer can store values from zero up to 255. Using a 16-bit string, the computer can store values up to 65,535.

Once an audio signal has been converted to digital, the waveform can be stored, read, and manipulated by the computer. The digital audio file is comprised of individual measured values (samples) stored as strings of binary digits.

The Analog-to-Digital Conversion Process

Converting an analog signal to digital involves two critical parameters: the sample rate and the bit depth.

Sample Rate. The sample rate refers to the frequency at which the incoming electrical signal is measured by the audio interface. The sample rate must be high enough to accurately represent the original analog signal.

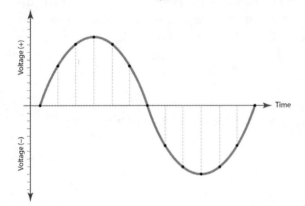

Figure 3.7 Sample intervals used to measure the changing voltage of an incoming audio signal

The sample rate required to accurately record a signal can be determined using a fundamental law of analog-to-digital conversion, known as the *Nyquist Theorem*. The Nyquist Theorem states that, in order to represent a given audio frequency, each cycle of the waveform must be sampled at least two times. Stated another way, the sample rate must be at least two times the frequency of the audio you wish to record.

Because we generally want to record full-spectrum audio – up to 20,000 Hz – the Nyquist Theorem tells us we need to use a sample rate of at least 40,000 Hz (or 40kHz). This will capture frequencies across the full range of human hearing.

 Most modern digital audio systems use a minimum sample rate of 44.1kHz.

Bit Depth. The bit depth (or word length) refers to the number of binary digits used to measure the voltage for each sample. The more binary digits used, the more accurate the measurements will be. Low bit-depth recordings exhibit a loss of audio detail—the result of a reduced dynamic range.

> ### What Is Dynamic Range?
>
> Dynamic range is the difference between the quietest signal that a system can represent and the loudest signal that the system can represent. Dynamic range is measured in decibels (dB). This is a relative measurement, not an absolute loudness value. The maximum absolute loudness of a system depends on the amount of amplification applied at the output; however, the dynamic range remains constant at all loudness levels, since it measures the relative *difference* between two levels.

You can estimate the dynamic range of a system by multiplying the bit depth by six. Said another way, each binary digit that is added to the word length increases the dynamic range by approximately 6 dB.

The useful amplitude range for speech and music extends from around 40 dB (very quiet) to around 105 dB (very loud): a 65-dB dynamic range. In order to produce that dynamic range, a digital audio system must use at least 11 bits in its word length (65 dB divided by 6 dB per bit). Additional dynamic range is necessary to provide headroom before clipping and to accommodate the noise floor inherent in a system. Addressing these considerations requires using more bits per sample.

 Most modern digital audio systems record with a minimum bit depth of 16 bits.

The Audio Interface

As mentioned previously, an audio interface is a device used to provide analog-to-digital conversion for the audio coming into a system. It also provides digital-to-analog conversion for the audio playing out of the system. The audio interface is what allows you to get sound into and out of your computer.

Most audio interfaces also provide one or more microphone pre-amps. Pre-amps generally include a gain control that allows you to adjust the incoming signal and optimize record levels.

Audio Interface Considerations

A multitude of audio interfaces are available today, providing an enormous array of options. To narrow down your choices, it helps to decide on the characteristics that are important for your recording needs.

Some factors to consider include:

- What kind of input and output (I/O) connectivity do you need?

- What level of sound quality are you looking for?
- What sort of budget are you working with?

I/O Connectivity

For the greatest flexibility, look for an audio interface with microphone (XLR) inputs.

If you are operating on a limited budget, you can consider an audio interface that includes just 1 or 2 microphone inputs. These may be supplemented with additional line- or instrument-level inputs. Line/instrument inputs can be useful for recording from an electric guitar or keyboard connected directly to the audio interface along with your miked source(s).

Most audio interfaces will include balanced stereo outputs, via a pair of ¼-inch jacks, for connecting to studio monitors. Many also include a stereo headphone jack with volume control. A headphone output is useful for monitoring playback from your DAW while recording: monitoring through headphones instead of your studio speakers helps prevent the playback from bleeding into a live microphone.

Sound Quality

Generally speaking, the "quality" of a digital audio file is a factor of the sample rate and bit depth used for the recording. But the recorded results are also influenced by the quality of the microphone pre-amps and the analog-to-digital (or A/D) converters in the audio interface.

Other factors also have a significant impact, including the recording environment and the quality and type of microphone(s) being used to capture the audio.

For these reasons, you should not consider the specifications of sample rate and bit depth in isolation when seeking to optimize the quality of your recordings.

When to Use High-Resolution Audio

For a simple recording project, you may find yourself using an inexpensive microphone to record audio that has a limited dynamic range (such as dialog for a podcast or webinar). You may also be recording in an untreated room or office environment. In such cases, recording with a high sample rate and bit depth will have no beneficial impact: recording at 44.1kHz and 16-bit will be sufficient.

Other times, you may be recording something with a wider dynamic range, such as a musical performance. This type of recording is often done using higher-quality microphones in a home-studio environment with some degree of acoustic treatment. In this case, you might consider using higher-resolution audio for better results and more editing flexibility. Recording at 24-bits is recommended in this situation to preserve the quality and dynamic range of the audio being captured.

> In cases where you find yourself working in a professional studio environment—with top-of-line microphones, boutique pre-amps, and other high-end gear—you will almost certainly want to record at a sample rate of 96kHz or above (and 24-bit) to capture all the subtle nuances of the performance and the character of the associated equipment.

Nearly all of the audio interfaces on the market today support sample rates up to 96kHz, with many supporting up to 192kHz. Likewise, 24-bit A/D conversion is a near universal standard for bit depth. This means that your search for an audio interface for Logic Pro X will not need to be based on the supported sample rate and bit depth of the device, as just about any device will fit the bill.

 Logic Pro X supports audio with sample rates up to 192kHz.

Instead, look for an audio interface with a construction quality and price point that complements your recording goals and budget. For higher-quality audio, look for devices from "brand name" companies that have a reputation for high-end gear…but expect to pay a premium for them.

Budget

The ultimate deciding factor often comes down to budget. But you can weigh the options, such as I/O connectivity and brand reputation, against your budget to find the right compromise.

For example, if you need only a single mike input, you may opt for a high-end audio interface to maximize the sound quality. On the other hand, if having 4 mike inputs is an absolute requirement, but sound quality is less of a concern, you might select a more utilitarian audio interface instead to stay within budget.

In the end, be sure to consider the budget for your entire recording setup as a whole. You will want to balance the cost of your audio interface against the other considerations covered in this chapter to ensure that you are not spending too much in one area and neglecting others.

Working Without an Audio Interface

In some cases, you may decide to work without an audio interface. You could do this as a temporary measure until you can save up for the audio interface you want. You could also do this as way to run your DAW when you are away from your audio interface, such as when on the road with a laptop.

In either case, Logic Pro X allows you to utilize the built-in audio on your computer (if available) in lieu of a dedicated audio interface. Launching Logic Pro X without an interface attached will automatically connect the available inputs and outputs on your computer to the software.

Review/Discussion Questions

1. When a sound event occurs, what are the changes in air pressure called (i.e., the areas where air pressure increases and decreases)? (See "The Basics of Audio" beginning on page 58.)

2. What is meant by the frequency of an audio waveform? How does the frequency of a sound wave affect our perception of the sound? (See "Frequency" beginning on page 58).

3. What characteristic affects our perception of the loudness of a sound? How is loudness measured? (See "Amplitude" beginning on page 59.)

4. What is the purpose of a traditional microphone? What three basic types of microphones are commonly available? (See "Traditional Microphones" beginning on page 59.)

5. How are USB microphones different from traditional microphones? What is the advantage of using a USB microphone over a traditional microphone? (See "USB Microphones" beginning on page 60.)

6. What are the differences between omnidirectional microphones, bidirectional microphones, and unidirectional microphones? Which is the most common type of mike for recording isolated sound sources? (See "Polar Pattern" beginning on page 61.)

7. What are some accessories available to help you with mike placement, noise reduction, and sound isolation when recording? (See "Microphone Accessories" beginning on page 63.)

8. How is multi-track recording different from recording to a single audio file? What are some advantages of multi-track recording? (See "What Is Multi-Track Recording?" beginning on page 65.)

9. What two parameters are critical for the analog-to-digital conversion process? What minimum specifications are used for these parameters in modern digital audio systems? (See "The Analog-to-Digital Conversion Process" beginning on page 67.)

10. What is the purpose of an audio interface? (See "The Audio Interface" beginning on page 68.)

11. What are some factors to consider when selecting an audio interface for use with your DAW? (See "Audio Interface Considerations" beginning on page 68.)

12. How is working without an audio interface useful? What is required to run Logic Pro X without an interface? (See "Working Without an Audio Interface" beginning on page 70.)

 To review additional material from this chapter and prepare for certification, see the Logic Audio Production Basics Study Guide module available through the Elements|ED online learning platform at ElementsED.com.

Exercise 3

Selecting Your Audio Production Gear

🎧 Activity

In this exercise, you will define your audio production needs and select components for a home studio based around Logic Pro X software. By balancing your wants and needs against a defined budget, you will be able to determine which hardware and software options make sense for you.

🕒 Duration

This exercise should take approximately 10 minutes to complete.

🎯 Goals/Targets

- Identify a budget for your home studio
- Explore microphone options
- Explore audio interface options
- Explore speakers/monitoring options
- Consider other expenses
- Identify appropriate components to complete your system

Getting Started

To get started, you will create a list of needs and define an overall budget for your home studio setup. This budget should be sufficient to cover all aspects of your initial needs for basic audio production work. At the same time, you'll want to be careful to keep your budget realistic so that you can afford the upfront investment. Keep in mind that you can add to your basic setup over time to increase your capabilities.

Exercise 3

Use the table below to outline your basic requirements and to serve as a guide when you begin shopping for options. Place an **X** in the appropriate column for your expected needs in each row.

Do not include MIDI gear in this table, as we will address that separately.

 This exercise assumes that you own a compatible computer with built-in speakers for playback. Do not include a host computer in this table.

Function or Component	Not Required	Minimum Configuration	Intermediate Configuration
Audio Interface: Output Channels (Playback)	–	2 Outputs (for stereo playback)	4-8 Outputs (for stereo playback and outputs to external gear)
Audio Interface: Input Channels (Recording)	–	2 Inputs	4-8 Inputs
Output Device(s)	Built-in Computer Speakers	Headphones	Stereo Monitor Speakers
Input Device(s) (Microphones)	–	USB Microphone	XLR Microphone(s) (specify type and number)
Accessories and Other (List or Describe) (Stands, Soundproofing, Software Add-Ons, etc.)			

Available Budget for the Above: _____

Identifying Prices

Your next step is to begin identifying prices for equipment that will meet your needs. Using the requirements you identified above as a guide, do some Internet research at an online music retailer of your choice to identify appropriate options for each of the items listed in the table below. You may also want to browse some manufacturers' websites for more information.

Component	Manufacturer and Model	Unit Cost
Audio Interface		
Microphones		
Headphones		
Monitor Speakers		
Accessories/Other		
	TOTAL	

Finishing Up

To finalize your purchase decisions, compare the total in the table above to the budget you allocated. If you find that your budget is not sufficient to cover the total cost, you will need to determine which purchase items you can postpone or consider bundle options. On the other hand, if you have money left over in your budget, you can consider upgrade options.

Bundle Options

One way to save some money is to look for gear bundles. These options can be much more affordable than buying the same components separately. Following are some good examples of bundles from Focusrite:

- Scarlett Solo Studio
- Scarlett 2i2 Studio
- iTrack Solo Studio

Each of the above bundles includes a Focusrite audio interface, a large–diaphragm condenser microphone, and a pair of closed-back studio headphones. As an added bonus, each bundle also includes several plug-ins

For more information, check out the details on the Focusrite website:

- Scarlett interfaces: https://focusrite.com/scarlett
- iTrack Solo interface: https://focusrite.com/ios-audio-interface/itrack/itrack-solo

Chapter 4

MIDI Recording Concepts

...*What You Need to Record MIDI*...

Understanding MIDI is essential to producing music using a DAW. In this chapter, we take a look at some basic MIDI concepts that will get you started creating tracks. We begin by taking a quick look at the history of MIDI and how the MIDI protocol works. Next we check out some of the types of MIDI controllers that are currently available. Then we look at how to set up the controller to communicate with your DAW. This leads us into a discussion on the fundamental differences between MIDI and audio data. Finally, we take a quick look at how to begin working with virtual instruments.

◎ Learning Targets for This Chapter

- Understand the basic history of MIDI
- Learn about the MIDI protocol
- Recognize types of MIDI controllers
- Learn how to set up your controller and DAW
- Recognize the differences between MIDI and audio
- Begin tracking with virtual instruments

Key topics from this chapter are illustrated in the Logic Audio Production Basics Study Guide module available through the Elements|ED online learning platform. Sign up at ElementsED.com.

The term MIDI stands for Musical Instrument Digital Interface. MIDI is a protocol for connecting electronic musical instruments, computers, and other devices, allowing them to communicate with one another.

The MIDI standard was developed in the 1980s to allow musical performance information to be shared among and between devices. The standard provides specifications for describing musical events, such as note values and durations, allowing musical performances to be represented numerically.

A Brief History of MIDI

Before MIDI, there was really no standardized way to have synthesizers and other electronic musical instruments communicate with each other. Early voltage-controlled analog synthesizers by Bob Moog and Don Buchla (et al.) could send *voltage* information to each other. But they couldn't communicate more complex musical ideas, such as a discrete note value with all of its pertinent information (pitch, amplitude, duration, etc.).

Figure 4.1 A Buchla 200e analog modular synthesizer

Digital Control

Another major drawback to analog synthesizers was that individual sound configurations could not be saved; favorite sounds had to be rebuilt by hand to be reused. Practitioners would actually make little drawings that showed where cables were patched and the position of each knob. (See Figure 4.2.) Some would even take Polaroid pictures so they'd have a photograph of the patch. (Interestingly, this way of working is making a huge comeback with the current Eurorack synthesizer craze!)

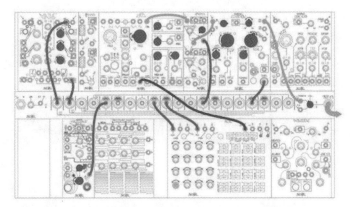

Figure 4.2 A patch "sheet" from a modern Eurorack analog synth

By the early 1980s, despite still using analog circuitry to generate sounds, most synthesizers had made the move to digital control for things like patch storage and recall (saving sounds and recalling them later). This created the possibility of having a common digital communication standard that would allow an electronic musical instrument to speak to any other instrument, regardless of the manufacturer.

The Birth of MIDI

In 1981, a proposal for a "Universal Synthesizer Interface" (USI) was presented at a meeting of the Audio Engineering Society (AES). This lead to the original MIDI 1.0 Specification. The specification was finalized in 1983 by a group that would become the MIDI Manufacturers' Association. The term *MIDI* was coined, as an acronym for Musical Instrument Digital Interface.

Figure 4.3 Sequential Circuits Prophet 600 (1982)

At the 1983 National Association of Music Merchants (NAMM) trade show, the first public unveiling of MIDI occurred. A MIDI connection was demonstrated between a Sequential Circuits Prophet 600 (shown in Figure 4.3) and a Roland Jupiter 6. Soon, every major manufacturer had added MIDI connectivity to their synthesizers, drum machines, sequencers, and other devices.

The MIDI standard also helped pave the way for computer-based music. In 1984, Passport Designs released their MIDI/4 sequencer that ran on a personal computer.

The MIDI Protocol

The complete MIDI protocol is quite complicated, but it can be informative to take a look at the basics. MIDI messages can be divided into three basic categories: MIDI notes, program changes, and controller messages.

MIDI Notes

MIDI notes are probably the most important part of the protocol. Note data is the building block for almost all the work you'll do with MIDI.

Note Values

One thing to understand about MIDI messages is that almost all parameters have a range of values spanning from 0 to 127. So, for instance, the pitch of a MIDI note is represented using a range of 0 to 127. A value of 0 represents the "C" note two octaves below the lowest note of a piano. We refer to this note as C-2.

Middle C is typically assigned a note value of 60 and referred to as C3 (although some manufacturers use C4 instead). Figure 4.4 provides a diagram of note values.

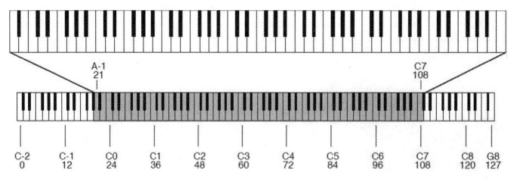

Figure 4.4 A standard piano keyboard with note names and MIDI note numbers

Note Velocities

The other important bit of note information is the velocity, or how hard a note is played. Velocity is also represented using a range of 0-127, with the value of 64 falling right in the middle (sort of a *mezzo forte* in musical terms). Playing a note lighter will result in a lower velocity value, whereas playing harder will result in a higher value.

Note On and Note Off Events

So, a MIDI note is a combination of a note number (pitch) and a velocity number. We refer to this combination as a **NOTE ON** command, because it activates the note in a technical sense. The Note On is eventually followed by a **NOTE OFF** command, which tells the device to stop playing the note. The Note Off command is the same note number, paired with a velocity of zero. The Note On and Note Off events are really all we need to send a note between two MIDI devices.

Logic Pro X provides an "Event List" that shows the note information for a track. From left to right, this list displays the bar and beat location where each note starts, the status, MIDI channel, the associated note number (pitch), the attack velocity, and the note length.

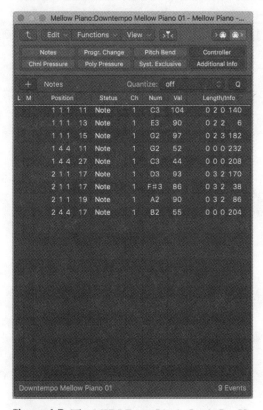

Figure 4.5 The MIDI Event List in Logic Pro X

Program Changes

As mentioned above, patch storage and recall was an early advantage of digital control for synthesizers. The MIDI protocol improved upon this, enabling patch change commands (also known as program changes) to be sent to and from devices.

Like other MIDI messages, patch change commands use a range of values from 0 to 127. Thus, most banks of sounds on a MIDI keyboard will have 128 sounds. Table 4.1 shows an example bank of 128 sounds that many synthesizers support. This is part of the *General MIDI* standard.

Table 4.1 General MIDI Instrument Families

Program Change #	Family Name	Program Change #	Family Name
1 – 8	Piano	65 – 72	Reed
9 – 16	Chromatic Percussion	73 – 80	Pipe
17 – 24	Organ	81 – 88	Synth Lead
25 – 32	Guitar	89 – 96	Synth Pad
33 – 40	Bass	97 – 104	Synth Effects
41 – 48	Strings	105 – 112	Ethnic
49 – 56	Ensemble	113 – 120	Percussive
57 – 64	Brass	121 – 128	Sound Effects

Controller Messages

The final category of MIDI messages we'll consider is the category of controller messages—also known as MIDI continuous control (CC) messages. These are messages that can be used for purposes such as capturing the expressiveness of a MIDI performance (pitch bend or mod wheel movements) or performing studio mixing tasks (volume or pan changes). MIDI CC data represents an essential part of how keyboardists actually *play* their instrument.

Table 4.2 provides a list of standard MIDI continuous control defaults. Notice that certain controller numbers are "undefined," leaving room to make custom assignments.

Table 4.2 MIDI Controller Numbers (Excerpt)

Controller Number	Hex	Controller Name
0	00h	Bank Select
1	01h	Mod Wheel
2	02h	Breath Controller
3	03h	Undefined
4	04h	Foot Controller
5	05h	Portamento Time
6	06h	Data Entry MSB
7	07h	Main Volume
8	08h	Balance
9	09h	Undefined
10	0Ah	Pan

MIDI Controllers

To get started working with MIDI, it helps to have a MIDI controller. A MIDI controller is a device that generates MIDI performance data and transmits it to another device, such as a DAW, for processing. You can use a MIDI controller for recording MIDI notes and performance data into your DAW or for triggering a sound module or virtual instrument.

A variety of different MIDI controllers are available today. The most common designs are based on the piano. With origins in the 18th century, the familiar black and white piano keyboard is a staple in the MIDI controller market. However, the last decade or so has seen an explosion in MIDI controller innovation. Today's choices include drumpad controllers (inspired by the drum machine designs of the 1980s), grid controllers (originally introduced to augment the grid-based interface of Ableton Live), and mixer controllers (borrowing their faders and knobs from established analog and digital mixer designs).

Tone-Generating Keyboards

Many keyboard-based devices on the market can be classified as *tone-generating keyboards*. These are keyboards capable of generating their own sounds. Although generally designed for stand-alone performance, most tone-generating keyboards also provide MIDI connectivity. This allows you to use the keyboard to generate MIDI performance data and record it on your DAW. It also allows the keyboard to receive MIDI data from elsewhere and use it to trigger the device's onboard sounds.

Tone-generating keyboards generally fall into two categories: digital pianos and synthesizers:

- **Digital Pianos**—Digital pianos are exactly what the name implies: they look like a scaled-down version of a real piano. Digital pianos typically feature a full 88-key keyboard (often with simulated piano action known as *hammer action*). These keyboards are designed to compete head-to-head with real acoustic pianos; thus they are very simple to operate.

    Digital pianos generally offer a limited range of sounds (such as piano and organ). However, more expensive models may include additional sounds and auto-accompaniment features that can generate bass lines and drum patterns. Perhaps the most distinguishing feature of a digital piano is the presence of built-in speakers, which synthesizers rarely offer. (See below.)

    Figure 4.6 Yamaha Arius Digital Piano

- **Synthesizers**—The term "synthesizer" encompasses a wide range of vintage and modern instruments. The term was originally used to describe analog devices that employed a combination of oscillators and filters to create sounds. These devices generated approximations of acoustic instruments (such as flute or violin) as well as otherworldly sounds that didn't bear resemblance to traditional instruments.

    But synthesizers quickly evolved to encompass both analog and digital technologies, with newer incarnations capable of using older analog-style synthesis as well as modern digital techniques such as sampling, physical modeling, and frequency modulation (FM) synthesis.

    Modern synthesizers come in a variety of sizes, ranging from small, highly-portable two-octave (25-key) models to full 88-key piano–sized models.

Figure 4.7 Moog's classic MiniMoog synthesizer (1970)

If you already own a digital piano or synthesizer, look for a MIDI 5-pin connector or USB port for connecting it to your DAW. If you don't already own such a device, you can consider using a keyboard controller instead.

Keyboard Controllers

Keyboard controllers are essentially piano-style controllers that provide MIDI output. You can use a keyboard controller to create or record a MIDI performance. However, it will not include any onboard sound or tone-generating capabilities. This means that you cannot use the controller as a stand-alone instrument. Instead, the keyboard is designed to control other hardware synthesizers, sound modules, or virtual instruments.

The keyboard controller serves as a kind of universal MIDI input device. Using a piano keyboard as the performance interface, keyboard controllers enable any user with basic piano skills to create just about any kind of MIDI performance with little to no added learning curve.

Keyboard controllers come in a variety of sizes and price points. Users on a limited budget can consider a basic 25- to 37-key controller for under a hundred dollars. More sophisticated models with up to 88 keys can be found in the hundred-fifty to five hundred dollar range. These commonly include knobs, faders, and drumpads that can be mapped to control almost any function of the target instrument.

Figure 4.8 M-Audio Oxygen 49 keyboard controller

Drumpad Controllers

Most modern drumpad controllers are descended from Roger Linn's ingenious designs from the 1980s. The most iconic of Linn's designs was the Akai Professional MPC60, introduced in 1988. The 4×4 layout of pads that Linn pioneered has since become a standard pad arrangement for dozens of products.

The pads themselves are carefully designed to have a great feel that helps the player to perform dynamic drum and percussion grooves.

Figure 4.9 Akai Professional MPD218 drumpad controller

Grid Controllers

Today, the grid controller is probably the most popular controller type after the piano-style keyboard. Grid controllers have become popular for studio work, but they have also been embraced by live performers in a variety of genres, including EDM and hip-hop.

The grid controller design was inspired by Ableton Live's session view, with each pad used to trigger clips, play drum sounds, and adjust controls such as volume. Some grid controllers, such as Novation's Launchpad Pro and Ableton's Push 1 & 2, can be used to enter notes as well. These include options for restricting notes to a musical scale so that no wrong notes can be played! And, while grid controllers are generally optimized to work with Ableton Live, they can be used with an increasing number of DAWs.

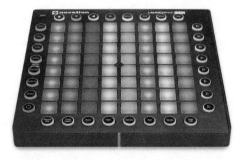

Figure 4.10 Novation Launchpad Pro grid controller

Alternate Controllers

You can also find some very innovative new controller styles on the market these days, offering alternatives to the standard keyboard, pad controller, or grid controller.

A primary focus of such alternative controllers currently is MIDI Polyphonic Expression (MPE). MPE is a new industry-wide standard that uses a separate MIDI channel for each touch. This channel-per-note configuration permits the controller to transmit discrete vibrato, glissando (pitch slides), note pressure, and other expressive information for each note that is played.

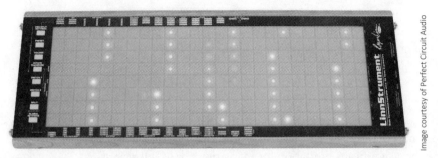

Figure 4.11 Roger Linn's Linnstrument is one example of an alternate controller

What to Look for in a MIDI Controller

Deciding which controller or controllers to purchase can be an overwhelming task! The most important thing is to determine which activities you will be performing frequently, which you'll be doing only occasionally, and which you won't be doing at all. Here are some suggestions based on the strength of various controller types for typical music production tasks:

- Entering Notes:
  - Keyboard (best choice)
  - Grid (good alternative)
- Playing Beats:
  - Drumpad (best choice)
  - Grid (good alternative)
  - Keyboard (also good)
- Arranging:
  - Grid

Purchasing a Keyboard Controller

If you decide to purchase a keyboard controller, you'll need to decide on a few other attributes. The most important is probably the number of keys (note range). Your needs here will generally be determined by your proficiency as a keyboard player.

If you are an accomplished keyboardist who plays with two hands simultaneously, you will probably feel most comfortable with a keyboard featuring 61 keys or more. A full-sized, 88-key controller will offer the widest note range and the best feel. On the other hand, if you play using just one hand at a time, you may be satisfied with a smaller 25-key model.

Aside from note range, you'll also want to consider a device's support for velocity sensitivity, aftertouch, pitch and mod wheel controls, and other mappable controls like knobs and sliders.

Velocity Sensitivity

As mentioned previously, representing how hard each note is played is often a key ingredient to the expressiveness of a performance. Most moderately sophisticated keyboard controllers will be velocity sensitive, meaning they measure how hard each key is struck. However, some budget models may not include this capability, which can be a significant limitation for many types of MIDI recording.

Aftertouch

Aftertouch lets a performer add expressive inflections to a sustained note or chord. Aftertouch can be used to add vibrato, pitch bends, and swells by varying the pressure on the held keys. When using a virtual instrument that supports it, aftertouch can be essential to making the instrument sound realistic. (Think of guitar solos and horn parts, for example: how often are sustained notes completely static?) Here again, budget keyboard controllers often do not include support for this parameter. Be prepared to do some comparison-shopping when trying to get the best value for your buck.

Pitch and Mod Wheel Controls

These are additional options for adding expressiveness to a performance. Pitch bend and mod wheel controls are common on synthesizers but may not be included on some MIDI controllers. A traditional pitch wheel adds pitch bend (portamento) control, allowing the player to bend a note (or chord) up or down in a continuously variable manner. Pitch bend controls are sometimes included in non-wheel format, such as in a joystick, knob, or ribbon format.

Modulation wheels (or mod wheels for short) are typically used to set a vibrato amount. However, the mod wheel can also be mapped to other parameters, such as volume swells, tremolo, and filter sweeps.

Other Mappable Controls

Additional buttons, knobs, and sliders may be included on mid-sized and full-sized keyboard controllers. These controls can be mapped to plug-in parameters and other MIDI controls. In some cases, they can even operate functions in your DAW, such as transport controls and volume faders.

Setup and Signal Flow

Once you've selected a MIDI controller, you'll need to establish communication between the controller and your computer. Almost all modern MIDI controllers use USB cables rather than MIDI cables to communicate. Configuring such a device can be as simple as plug-and-play, if the device is USB class-compliant. However, some devices require that you install specific drivers built for your computer and operating system.

Plug-and-Play Setup

For "class-compliant" USB devices, no drivers are required to communicate with your computer. You can verify that a device is class-compliant by checking the specifications on the manufacturer's website. Alternatively, you can connect the device to your computer and launch the appropriate software utility to verify communication.

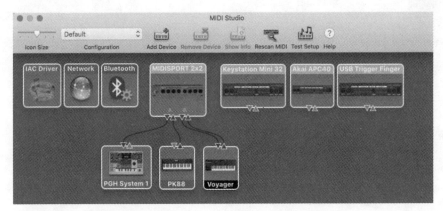

Figure 4.12 An example MIDI setup in Apple's Audio MIDI Setup utility; class-compliant devices will appear here automatically

MIDI Cables and Jacks

If you are using a device that has regular MIDI connections (not USB), you will have to deal with MIDI cables and jacks. A MIDI cable is technically called a 5-pin DIN cable and has connectors as shown in Figure 4.13.

Figure 4.13 A pair of MIDI cables

MIDI data is actually transmitted using only two pins on the 5-pin connector.

An important aspect of a MIDI cable is that it can carry *16 channels* of information. That means you could have up to 16 patches playing on one synthesizer, and they could all be addressed individually using a single MIDI cable.

The jacks that MIDI cables plug into are typically called *ports*. These can include an input, an output, and sometimes a "thru" connection.

Figure 4.14 A typical configuration of MIDI ports on a device

The MIDI In port receives incoming MIDI data from an outside source; this is commonly used to play back a MIDI performance coming from a DAW or other source using the onboard sounds of the device. The MIDI Out port sends MIDI data out from the device, for recording to a DAW or for triggering a different sound module. The MIDI Thru port passes the signal from the In port directly out to another device, without modifying it by any performance being done on the device. This can be very handy if you have a small MIDI interface with only a couple of ports, or if you have a complex live performance setup and don't want to lug an interface around.

Using a MIDI Interface

Although most modern controllers use USB, you may encounter a hardware device (such as an older synthesizer) that uses traditional 5-pin MIDI cables to communicate. If so, you'll need a way to connect the MIDI cables to your computer, because the computer will not have the appropriate jacks for MIDI. You can accommodate these connections using an audio interface with MIDI jacks (quite common on all but the smallest interfaces).

Figure 4.15 MIDI jacks on the back of a Focusrite 2i4 audio interface

As an alternative to an audio interface, you can use a dedicated MIDI interface to route MIDI data to your computer. MIDI interfaces come in a variety of shapes and sizes and can have as many as eight jacks for MIDI input and output.

Figure 4.16 M-Audio Uno MIDI interface

Considerations for Using Multiple MIDI Devices

Even the smallest studio will often have more than one MIDI device. This could entail any combination of keyboard, drumpad, grid, and mixer controllers. Aside from the ergonomic issues of placing multiple controllers in the physical space, you'll also need to consider whether you want the devices to control specific instruments or to all control the currently selected instrument.

In most DAWs, the ideal solution is to leave inputs on your MIDI tracks assigned to "ALL" so that the currently selected instrument will respond to any controller that you touch. This makes it very easy to play keyboard for a piano part, then switch over to the drumpads to perform a beat, all without having to stop and change the MIDI routing.

MIDI Versus Audio

One of the more difficult concepts to grasp when you begin working with a DAW is the difference between MIDI and audio. While most novice audio producers have a decent understanding of how audio data is recorded and represented in a DAW, many do not have a similar understanding with regard to MIDI data.

What Is MIDI?

As discussed earlier in this chapter, MIDI is a protocol that facilitates communication between a huge range of hardware and software devices. MIDI data is not audio, but rather a series of messages that can communicate information such as notes (pitch), duration, velocity (how hard a key or pad is pressed), dynamics (volume), and more. A tone-generating hardware or software instrument can receive MIDI messages and convert them into audio data.

In other words, the MIDI data has the potential to become music that we can hear, but there's no way to listen to the MIDI data by itself.

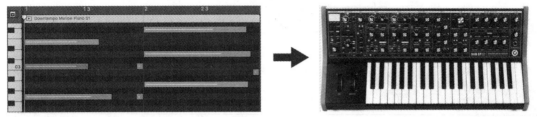

Figure 4.17 MIDI data (in piano roll format) must be sent to a tone-generating device to become music that we can hear

MIDI messages share some features with traditional music notation. Both contain the information necessary to play a piece of music, but neither is capable of actually becoming music on its own. Like MIDI, the notes of a musical score aren't audio, but instead represent the potential for audio once they are played using an instrument.

MIDI and notation contain information that can convey essential musical elements such as pitch, duration, velocity, and dynamics. Almost every DAW allows MIDI data to be viewed and edited as music notation (in addition to piano roll, event list, and other more modern formats).

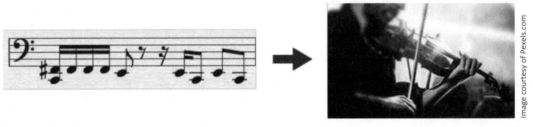

Figure 4.18 Music notation must be read by a musician to become music we can hear.

Monitoring with Onboard Sound Versus Virtual Instruments

If you are using a tone-generating keyboard such as a digital piano or synthesizer, you will need to decide whether you want to monitor from the onboard sound of the device or, alternatively, silence the onboard sounds and instead listen to a virtual instrument inside of your DAW.

It is rare that you will want to hear the sound of the controller while recording MIDI data into your DAW. (In fact, this can result in an annoying doubling of the controller audio with the DAW audio.) In most situations, you'll want to monitor from the associated virtual instrument's output alone. This can be accomplished by turning off "Local Control" on the keyboard device. This setting will disable the internal sounds of the device while still sending MIDI data out to the DAW.

 Refer to the user guide or manufacturer's website to find the local control setting on your particular device.

Tracking with Virtual Instruments

Most modern music production studios incorporate software instruments, or virtual instruments, into their toolkit. In fact, a number of software manufacturers are known primarily for their virtual instrument offerings, including Native Instruments, Arturia, and Spectrasonics. Virtual instruments encompass a huge range of musical devices, including digital synthesizers, analog-modeling synthesizers, samplers, drum machines, and much more. They have become so ubiquitous in the DAW world that every major manufacturer offers a range of free or paid virtual instruments.

Figure 4.19 Native Instruments Kontakt sampler

Creating Tracks for Virtual Instruments

Working with virtual instruments in your DAW can seem a bit confusing at first. It's not quite as simple as just creating a track and pressing record! The first step to working with virtual instruments is to create the appropriate track type, which is usually referred to as an "Instrument" or "MIDI" track depending on the DAW. These tracks are capable of recording and editing MIDI data, routing the MIDI data to a virtual instrument, and then mixing the virtual instrument's resulting audio output.

Let's take a look at a couple of example configurations.

Virtual Instruments in Logic Pro X

Routing MIDI to a virtual instrument in Logic Pro X is quite similar to the process in many other DAWs, including Pro Tools, Cubase, and others.

94 Chapter 4

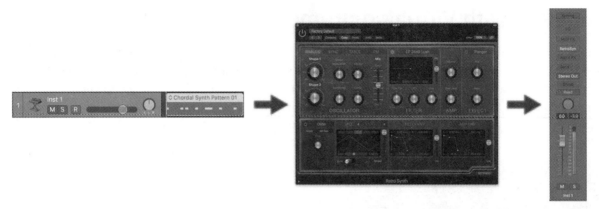

Figure 4.20 When MIDI is routed to the Retro Synth plug-in, the plug-in's audio output is processed in the track's Fader section.

In Figure 4.20, the **Inst 1** track sends its MIDI data to the **Retro Synth** software instrument plug-in, which is inserted on the same track. Then, Retro Synth converts the MIDI data to audio, with the sound being determined by the current patch/preset. Next, the audio output from Retro Synth is routed through the audio section of the **Inst 1** track, where its volume, pan, and other audio attributes can be adjusted as desired. Finally, the audio output of the track is routed to the audio interface, so that it can be monitored through speakers or headphones.

Virtual Instruments in GarageBand

GarageBand provides another example of this concept, although the approach used in the software is somewhat different. (See Figure 4.21.) Note that GarageBand doesn't offer a mixer view, so the track is only visible in a horizontal layout that is similar to the Main window in Logic Pro X.

In GarageBand, the sound module is assigned by selecting the desired source from the Sound Library, rather than by placing a virtual instrument plug-in on the track.

Figure 4.21 Using a drum instrument in GarageBand

In Figure 4.21, we can see the MIDI data in the familiar piano roll format on the track. That MIDI data is being routed to the East Bay drum kit (whose controls are displayed at the bottom). The kit converts the MIDI data to audio. The audio output actually appears on the horizontal controls just to the left of the MIDI data, where the volume and pan attributes can be modified. From there, the audio goes to the interface output.

Summary

The function of a virtual instrument is quite similar regardless of the DAW you choose. Professional-level DAWs offer sophisticated routing and additional views for working with MIDI data (as well as audio). By contrast, entry-level DAW options (such as GarageBand) aim to keep things as simple as possible. Regardless of where you begin, once you have gained basic familiarity with your chosen DAW, you'll find it relatively easy to apply that knowledge to any other DAW you should encounter.

Review/Discussion Questions

1. What were some of the drawbacks of historical analog synthesizers? (See "A Brief History of MIDI" starting on page 78.)

2. How did digital control of synthesizers pave the way for the development of MIDI? (See "A Brief History of MIDI" starting on page 78.)

3. What are the three basic categories of MIDI messages? (See "The MIDI Protocol" starting on page 80.)

4. What is the standard range of values for all MIDI messages? (See "Note Values" starting on page 80.)

5. What MIDI note number is typically used to represent middle C on a piano? Is this number the same with every device? (See "Note Values" starting on page 80.)

6. What are the two components of a MIDI Note On command? (See "Note On and Note Off Events" starting on page 81.)

7. What is another name for a patch change command when using MIDI? (See "Program Changes" starting on page 81.)

8. What type of MIDI data can be used to increase the expressiveness of a MIDI performance? (See "Controller Messages" starting on page 82.)

9. Aside from keyboards, what are some other types of MIDI controllers? (See "MIDI Controllers" starting on page 83.)

10. What are two categories of tone-generating keyboards? (See "Tone-Generating Keyboards" starting on page 83.)

11. What does it mean when a USB device is labeled class-compliant? What is the benefit of a class-compliant USB device? (See "Plug-and-Play Setup" starting on page 89.)

12. How can MIDI notes be converted to an audio signal that we can hear? (See "MIDI Versus Audio" starting on page 91.)

13. How can a virtual instrument plug-in be assigned in Logic Pro X? (See "Tracking with Virtual Instruments" starting on page 93.)

To review additional material from this chapter and prepare for certification, see the Logic Audio Production Basics Study Guide module available through the Elements|ED online learning platform at ElementsED.com.

Exercise 4

Selecting Your MIDI Production Gear

🎧 Activity

In this exercise, you will define your MIDI production needs and select components to complement the Logic Pro X system you configured in Exercise 3. By identifying the type of MIDI production work you will be doing, you will be able to select an appropriate MIDI controller (or controllers) to meet your needs.

⏱ Duration

This exercise should take approximately 10 minutes to complete.

⊕ Goals/Targets

- Identify a budget for your MIDI hardware
- Explore controller options
- Consider other expenses
- Identify appropriate components to complete your system

Getting Started

To get started, you will create a list of requirements for your MIDI controller and define an overall budget for your MIDI setup. Once again, make sure your budget is sufficient to cover all your immediate needs, while remaining realistic about what you can afford. You can always add to your basic MIDI setup over time with more devices and virtual instruments.

Use the table below to identify your basic MIDI requirements and to serve as a guide when you begin shopping for a controller. Place an **X** in the appropriate column(s) for each row.

Criteria	Rating			
Piano Keyboard Skills	Novice / Hunt and Peck	Competent Beginner	Intermediate	Advanced
Intended Uses	Drum Programming, Effects, Short Musical Parts, Loops, EDM	Background Music Production, Simple Keyboard Parts and Other Instruments	Lead Instrument Parts, 2-Handed Performances, Complex Arrangements	Orchestral Scores and Arrangements
Feature Requirements	Basic MIDI Input	Velocity Sensitivity	Pitch Bend and Mod Wheel	Aftertouch
Accessories and Other (List or Describe) (Keyboard Stands, Foot Pedals, etc.)				

If your **X**'s fall predominantly in the left half of this table, you may want to consider a drum pad or grid controller, or look for a budget-model keyboard controller. If your **X**'s fall predominantly in the right half, you should probably target a full-sized and fully featured keyboard controller. If your MIDI needs are extensive, you may want to consider selecting more than one type of controller to purchase.

Available Budget for MIDI Gear: _____

Identifying Prices

Your next step is to begin identifying prices for the type of controller and accessories that will meet your needs. Using the requirements you identified above as a guide, conduct some research online to identify available options. List potential matches in the table below, along with short descriptions (including number of keys, if applicable) and prices.

Type of Controller	Manufacturer and Model/Description	Price
Accessories/Other		

Total Cost for MIDI Gear: _____

Finishing Up

To finalize your purchase decisions, total the cost of each option plus your required accessories and compare this to the budget you allocated. If you find that your budget is not sufficient to cover your preferred choice, consider a different option with the plan to upgrade in the future.

If you are able to purchase your first choice items and have money left over in your budget, you can look at software add-ons to supplement your collection of virtual instruments.

Chapter 5

Logic Pro Concepts, Part 1

...What You Need to Know to Get Started with Logic Pro X...

This chapter introduces you to some basic operations and functions you need to be familiar with to get started working in Logic Pro X. We cover how to create a new project, how to access and use the main windows, and how to create tracks. We also cover basic navigation and selection techniques, including transport controls, zooming functions, and scrolling operations. These foundational concepts will help you get up and running and remain productive as you work.

◈ Learning Targets for This Chapter

- Learn how to create a new Logic Pro X project
- Explore the main window and its sections
- Learn how to create tracks and understand the supported track types in Logic Pro X
- Understand the difference between mono tracks and stereo tracks and recognize when to use each
- Learn how to navigate in the main window by scrolling and zooming
- Learn how to browse for media and add loops to your project

 Key topics from this chapter are illustrated in the Logic Audio Production Basics Study Guide module available through the Elements|ED online learning platform. Sign up at ElementsED.com.

If you need professional-level audio production capabilities on a limited budget, Logic Pro X is an excellent choice. Logic Pro X has a simplified feature view for new users, allowing you to get up and running quickly. Once you're ready to dig a little deeper you can turn on Logic's Advanced Tools and get access to under the hood commands.

Creating a Project

When you open Logic Pro X for the first time, the first thing you see after startup completes is the Project Chooser. If you've used Logic previously, the last project you were working in will automatically open upon startup. In this case, you can close the file by selecting **FILE > CLOSE PROJECT** from the main menu bar. Then choose **FILE > NEW** or press **COMMAND+SHIFT+N** to display the Project Chooser. The Project Chooser is your springboard for creating a new Logic Pro X project or opening an existing project.

Figure 5.1 The Project Chooser in Logic Pro X

For this chapter, we will start by creating a new empty project.

To create a new empty project in Logic Pro X, do the following:

1. Display the Project Chooser, if not currently showing, by choosing **FILE > NEW FROM TEMPLATE**.

2. Verify that the **NEW PROJECT** option is selected in the left sidebar of the Project Chooser.

3. Select **EMPTY PROJECT** in the main window area.

4. Click the **CHOOSE** button at the bottom of the Project Chooser. The New Tracks dialog box will appear, allowing you to add tracks to the project.

> ⓘ You can also open an existing project from the Project Chooser by clicking on the **OPEN AN EXISTING PROJECT** button at the bottom left.

Using Project Templates

Logic includes several pre-made templates under the **PROJECT TEMPLATES** option in the left sidebar of the Project Chooser. These are starting points that open up with instruments and effects preconfigured. If you often use the same set of instruments, a template can help you avoid having to build the same complex project setup each time from scratch.

You can create your own templates from project files you have created. After building out a project, select **FILE > SAVE AS TEMPLATE**.

Selecting a Track Type

After creating a new empty project, the New Tracks dialog box will display. Logic requires at least one track to be present in a project at all times.

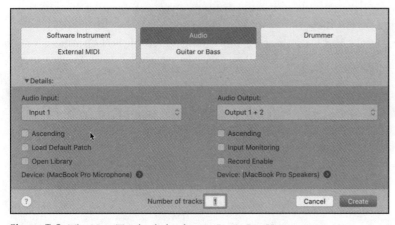

Figure 5.2 The New Tracks dialog box in Logic Pro X

The most common types of tracks you'll be working with for this course are Audio tracks and Software Instrument tracks. Logic provides five different track types:

- **Audio**—Audio tracks are used for working with different types of audio regions. You can import, record, edit, and arrange audio files and audio loops on an Audio track. If you plan to record on the track, you can select a specific input on your interface to assign when creating the track.

> (i) Regions are pieces of media that appear as rounded rectangles in the Tracks area. The two primary region types are audio regions, which show the audio waveform, and MIDI regions, which contain note events and other MIDI data.

- **Software Instrument**—Software Instrument tracks are used for working with MIDI regions and virtual instruments. You can import, record, edit, and arrange MIDI files and MIDI loops on a Software Instrument track. The MIDI data will trigger sounds from an internal virtual instrument audio effect.

- **External MIDI**—External MIDI tracks are used for working with external MIDI devices. As with a Software Instrument track, you can import, record, edit, and arrange MIDI data and MIDI loops on an External MIDI track. However, the track will output its MIDI data to an outside synthesizer, drum machine, or sound module to trigger sounds.

- **Guitar or Bass**—A Guitar or Bass track is an Audio track that is preconfigured with a plug-in amplifier to enhance the sound of an electric guitar or bass guitar connected directly to your audio interface. You can import, record, edit, and arrange audio files and audio loops on a Guitar or Bass track, just as with an Audio track.

- **Drummer**—A Drummer track provides pre-built adjustable drum grooves. A Drummer track is similar to a software instrument track, but it contains only *Drummer regions*, rather than MIDI regions. Parameters for Drummer tracks and Drummer regions can be edited in the Drummer Editor.

> (i) The Project Chooser is also used to specify parameters for your project, such as the sample rate and bit depth, selected interface, key and time signature, and project tempo. These options can be shown or hidden by clicking on the disclosure triangle to the left of the word Details.

Enabling Advanced Tools

Before starting work in Logic for this course, you will need to enable the Advanced Tools, as shown in Figure 5.3. To access these settings, select **Logic Pro X > Preferences > Advanced Tools**. For this course, you should check all of the available boxes, so that everything is enabled.

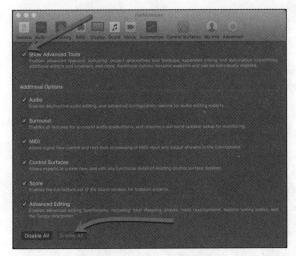

Figure 5.3 The Advanced Tools settings in Logic Pro X's Preferences

Main Window

When working in Logic Pro X, you'll spend most of your time in the main window. You can toggle and resize several panes to access all the tools and media you'll need for your project. The main window has three primary areas in Logic's default configuration: The control bar, the Inspector, and the Tracks area:

- **Control bar**—The control bar is displayed at the top of the main window. This area contains various buttons that activate panels in the display. It also includes transport controls for controlling project playback, an LCD area showing location information, and a Master volume slider to control overall output levels.

- **Inspector**—The inspector area on the left side of the main window provides different contextual options based on what track, region, or area is in key focus. The lower part of the inspector shows channel strips for the selected track and its associated output. You can use this area to view and edit channel strip parameters and plug-ins.

- **Tracks area**—The Tracks area is the central part of the main window. This is the primary workspace of Logic Pro. You will use this area to work with audio and MIDI material for your project. Audio waveforms and MIDI performances appear in a timeline display on your tracks in this part of the window.

Figure 5.4 The main window in Logic Pro X, with the Inspector open on the left side (default view)

The Control Bar

The control bar runs along the top of the main window. This area has several buttons to toggle other displays on or off. It also includes transport controls for playback operations and information about the project meter, tempo, and key signature.

Figure 5.5 The control bar in the Logic Pro X main window

The Library

The **LIBRARY** button on the control bar displays a panel that contains patches and FX chains for the selected channel strip or plug-in. Users can access this panel to quickly browse different presets and try out sounds. The Library panel can also be engaged/disengaged by pressing the **Y** key on the computer keyboard.

Figure 5.6 The Library button selected in Logic Pro's control bar

The Inspector

The **INSPECTOR** button opens the Inspector panel. The Inspector can also be engaged/disengaged by pressing the **I** key.

The image below shows the **Hi-Hat** track in the Inspector view. The right column typically displays track information for hardware outputs or Aux return tracks. Here the left column shows the channel strip for the **Hi-Hat**, while the right shows the settings for the **Stereo Output**.

Figure 5.7 The Inspector button selected in Logic Pro's control bar and the resulting display

The Quick Help Button

The **QUICK HELP** button displays coaching tips as your mouse hovers over areas and tools. Quick Help can also be engaged/disengaged by pressing **SHIFT+/**.

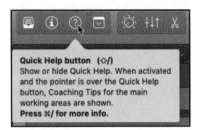

Figure 5.8 The Quick Help button selected in Logic Pro's control bar

The Toolbar

The Toolbar is an optional display that appears right below the control bar. Clicking the **TOOLBAR** button in the control bar will toggle this row of commonly used edit tools on/off. The Toolbar can also be engaged/disengaged by pressing **CONTROL+OPTION+COMMAND+T**.

Figure 5.9 The Toolbar button selected in Logic Pro's control bar

Figure 5.10 The Logic Pro Toolbar

 You can customize the display in the control bar or the Toolbar by Control-clicking and selecting CUSTOMIZE CONTROL BAR AND DISPLAY or CUSTOMIZE TOOLBAR, respectively, from the pop-up menu.

Smart Controls

The **SMART CONTROLS** button opens a panel that enables you to edit a track's sound in a limited capacity without having to open up the current channel strip or effects chain. The Smart Controls panel can be enabled/disabled by pressing the **B** key.

Logic Pro Concepts, Part 1   109

Figure 5.11 The Smart Controls button selected in Logic Pro's control bar

Mixer

The **MIXER** button opens up Logic Pro's Mixer display. In the Mixer, you can add and edit plug-ins, adjust volume and pan settings, and customize signal routing options. The Mixer's functions are discussed further later in this chapter. The Mixer display can be enabled/disabled by pressing the **X** key.

Figure 5.12 The Mixer button selected in Logic Pro's control bar

Editors

The **EDITORS** button opens a panel at the bottom of the main window, giving you access to either editing tools for audio tracks and regions or controls for the piano roll and score editor, based on what's selected. Editors can be enabled/disabled by pressing the **E** key.

Figure 5.13 The Editors button selected in Logic Pro's control bar

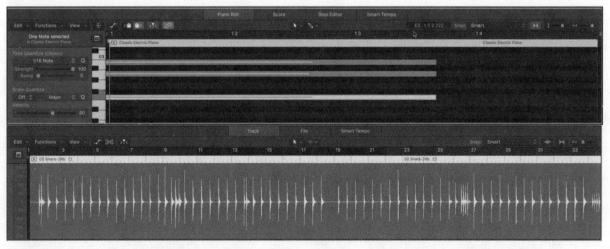

Figure 5.14 The open Editor panel. The piano view for a software instrument track (top) and the audio tracks view (bottom) are displayed in Logic Pro's main window.

Transport Controls

The Transport controls in the control bar provide buttons for various transport functions, such as play, stop, fast-forward, and rewind. These functions are available just to the left of the LCD in the main window. The transport controls can also be accessed in a floating Transport window. The floating Transport window is accessible by selecting **Window > Open Transport Float**.

Figure 5.15 Transport floating window in Logic Pro X

The LCD

The LCD in Logic Pro's control bar (and floating Transport window) displays a lot of helpful information about your project. You can double-click on any of the LCD sections to make changes for your project.

Figure 5.16 The LCD area in Logic Pro X

 You can also open a larger floating Bars and Beats or Timecode and minutes and seconds display from the display mode pop-up by selecting Open Giant Beats Display, and Open Giant Time Display respectively.

By default, the LCD is set to **Beats and Project**. This shows your project's playhead position, tempo, meter, and key signature. You can choose settings like **Beats and Time** to display the playhead location in a combination of musical and absolute time formats.

The Beats and Project LCD view displays the following information:

- **Playhead Position**—The left portion of the LCD indicates the position of the playhead, shown in Bars and Beats by default. It can also display smaller beat division and ticks, as well as absolute timecode/minutes and seconds values. You can click and drag on this display, or double-click and type in a new value, to reposition the playhead.

- **Tempo**—The center LCD section displays your project's tempo. This can be set to follow your project's set tempo in **Keep** mode, or to adapt to perceived tempo of recordings or imported audio with **Adapt** mode. A third mode, **Auto**, allows Logic to choose what best suits the situation. For this course, we will leave this set on **Keep**.

- **Project Information**—The right portion of the LCD shows information about your project, like its time signature and key signature. You can click on these values to change them.

 The LCD's current display can be changed by clicking on the display mode pop-up selector (down arrow) next to the project information section.

Replace

The **REPLACE** button lets you replace a recording or portion of an audio region during a record pass. The Replace button can be enabled/disabled by pressing the **/** key.

Figure 5.17 The Replace button selected in Logic Pro's control bar

Tuner

The **TUNER** button displays an onscreen tuner for any real instruments connected through your audio interface. You can use this feature to tune up an instrument prior to recording.

Figure 5.18 The Tuner button selected in Logic Pro's control bar

Solo

The **SOLO** button lets you listen to selected regions in isolation from the rest of your mix. Soloing can be enabled/disabled by pressing **CONTROL+S**.

Figure 5.19 The Solo button is selected in Logic Pro's control bar

Count In

The **COUNT IN** button is used in conjunction with Logic Pro's Metronome. The Count In feature gives performers a predetermined number of bars of click before recording starts. The Count In length can be specified by selecting **RECORD > COUNT IN** and then choosing the number of bars from the menu. Alternatively, you can select **FILE > PROJECT SETTINGS > RECORDING** and select the number of bars from the Count-in pop-up menu.

Count In function can also be enabled/disabled by pressing **SHIFT+K**.

Figure 5.20 The Count In button selected in Logic Pro's control bar

Metronome

The **METRONOME** button activates a steady click during recording and/or playback. The tempo of the metronome click is determined by the project tempo. The Metronome can be enabled/disabled by pressing the **K** key.

Figure 5.21 The Metronome button selected in Logic Pro's control bar

You can click and hold the metronome button to configure additional options from a pop-up menu.

Figure 5.22 Additional metronome options

Options available in this menu include the following:

- **Simple Mode**—In this mode, the metronome only sounds when you toggle it on and off. Other settings are unavailable.

- **Click While Recording**—Use this setting if you want the metronome to sound during record passes, whether it's toggled on or not.

- **Only During Count-In**—In this mode, the metronome will sound only during the count-in prior to recording. (Click While Recording must also be enabled for this to work.)

- **Click During Playback**—This setting causes the Metronome to sound during playback. This has the same effect as toggling the metronome on and off.

List Editors

The **LIST EDITORS** button shows a list of markers, MIDI events, time signature changes, and key signature changes. Here you can view or add events to any of these lists. The List Editors display can be enabled/disabled by pressing the **D** key.

Figure 5.23 The List Editors button selected in Logic Pro's control bar

Note Pads

The **NOTE PADS** button displays a panel that lets users create, view, and edit text notes for the entire project, or for specific tracks. Note Pads can be enabled/disabled by pressing **OPTION+COMMAND+P**.

Figure 5.24 The Note Pads button selected in Logic Pro's control bar

Loop Browser

The **LOOP BROWSER** button displays a tabbed panel that lets you browse and add pre-recorded Software Instrument (MIDI) and audio loops to your project. Logic includes a large library of loops. Each loop has information about key, tempo, and length. You can mark loops as favorites and browse using descriptors like genre or instrument type. You can drag loops onto existing tracks, or drag them into the workspace below your existing tracks to add them to a new track. The Loop Browser can be enabled/disabled by pressing the **O** key.

Figure 5.25 The Loop Browser button selected in Logic Pro's control bar

Browsers

The **BROWSERS** button displays the Browser panel. This panel lets you view any audio or media files that are currently in your project or search for media on your system. You can also import settings and data from other projects. The Browsers panel can be enabled/disabled by pressing the **F** key.

Figure 5.26 The Browsers button selected in Logic Pro's control bar

Tracks Area Menu Bar

Beneath the control bar in the main window is a thin area with additional controls and selectors known as the Tracks area menu bar. This area of the user interface contains menus that are specific to the Tracks area. It also provides tool menus, controls for showing track automation, the Catch button, Snap and Drag pop-up menus, a Waveform Zoom button, and scroll and zoom sliders.

Figure 5.27 Tracks area menu bar in the main window

 Many of the functions in the Tracks area menu bar are discussed later in this chapter and elsewhere in this book.

Rulers

Rulers are horizontal displays that appear at the top portion of the workspace area, right above your tracks.

Figure 5.28 Rulers appear below the Tracks area menu bar and above the regions in the Track area display.

Logic Pro has two views for the primary ruler: Musical, which shows time in bars and beats, and Absolute, which shows time in minutes and seconds. To display the Absolute ruler (minutes/seconds) as a secondary ruler, you can click on the View menu in the Tracks area menu bar and click SECONDARY RULER. Alternatively, you can use the shortcut COMMAND+OPTION+CONTROL+R.

Figure 5.29 The Musical Ruler displaying bars and beats in Logic Pro

> ⓘ To display only the Absolute ruler, you can select FILE > PROJECT SETTINGS > GENERAL and disable the Use Musical Grid checkbox.

You can position the playhead at a specific location by clicking in the lower portion of the ruler.

> ⓘ You can also position the playhead by SHIFT-CLICKING on a blank track area.

Mixer Display and Window

The Mixer display provides a mixer-like environment within the main window for recording and mixing audio. In this display, tracks appear as mixer strips. Each strip includes controls for inserts, sends, input and output assignments, panning, and volume.

Figure 5.30 The Mixer open in Logic Pro's main window

Remember that you can access the mixer when it is not displayed by clicking the Mixer button in the control bar or by pressing the **X** key.

 You can open a floating Mixer window by selecting Window > Open Mixer or by pressing Command+2.

The Mixer display and the floating Mixer window provide the following controls:

- **Signal Routing Controls**—The top portion of each track's mixer strip provides controls for routing signals for the track. These include Audio or MIDI FX selectors for applying plug-in processors, Send selectors for routing to an output or bus, an Input selector for routing audio into the track for recording or processing, and an Output selector for routing audio out of the track for playback. At the bottom of this portion of the channel strip is an automation mode selector.

 We discuss automation in Chapter 9 of this book.

Figure 5.31 Signal routing controls

- **Record and Playback Controls**—The bottom portion of the mixer's channel strip provides controls for setting record and playback options. These include Pan controls for positioning the signal within the stereo field; buttons to enable Input monitoring, Record, Solo, and Mute functions; and a Volume Fader for setting the output level. (See Figure 5.32.)

Figure 5.32 Record and playback controls

Working with Tracks

Once you've created a project in Logic Pro X, you will need to create tracks for your project. Tracks are where your audio and MIDI performances are recorded and edited. Audio and MIDI data can be copied and duplicated in different locations to create repeating patterns, to arrange song sections, or to assemble material from multiple takes. To work with the media on your tracks, you will need to select the appropriate tool for the work you will be doing.

Adding Tracks

To add tracks to your project, you can use the New Tracks dialog box, shown in Figure 5.2 earlier in this chapter.

To access the New Tracks dialog box, choose **TRACK > NEW TRACKS**, or press **COMMAND+OPTION+N**. Alternatively, you can click on the plus sign just above your tracks list.

Figure 5.33 The Add Tracks button in the main window

Mono Versus Stereo Tracks

Tracks that route or process audio signals offer a choice between mono and stereo channel formats. Either format will allow you to pan the track's playback in order to position the audio output within the stereo field. The distinction between them is the channel format of the track's *input* signal. In other words, the channel format you choose should be based on whether the incoming audio is mono or stereo.

For example, you will typically use mono audio tracks to record vocal and dialog performances, since the source signal will usually be a solo voice recorded with a single microphone. The same holds true for recording many types of close-miked acoustic instruments, as well as electric instruments that are connected directly to an audio interface. Common exceptions would include stereo-miked instruments, such as pianos, and electronic instruments that provide stereo outputs, such as drum machines and synthesizers.

In Logic Pro, you switch a track between mono and stereo after creating it. Audio tracks are created in mono format by default. You can change the format by clicking on the Channel Mode button in either the mixer or inspector windows. This button will appear as either a single circle (mono) or two linked circles (stereo). Clicking the Channel Mode button will toggle the track between the mono and stereo.

Figure 5.34 The Channel Mode button in the inspector on a stereo track (left) and in the mixer on a mono track (right)

External MIDI Versus Software Instrument Tracks

To record a MIDI performance, you can use either an External MIDI track or a Software Instrument track. External MIDI tracks are designed to route MIDI data out of Logic Pro to an outboard synthesizer or sound module. Software instrument tracks use virtual instruments included with Logic Pro or third party plug-in virtual instruments.

Both track types allow you to record incoming MIDI data from a connected controller. You can also edit MIDI notes, adjust note velocities, and record and edit MIDI CC data on either track type.

Configuring MIDI Tracks for Playback

During playback, an external MIDI track makes no sound on its own. In order for the notes on an External MIDI track to trigger audio, the MIDI output from the track must be routed to a hardware sound-generating device. For example, the MIDI playback can be routed out of Logic Pro to an external synthesizer to generate sound. The synthesizer sounds could then be monitored externally, by using an amplifier, or internally, by routing the synthesizer's audio back into the external MIDI track or into a record-enabled Audio track.

Configuring Software Instrument Tracks for Playback

In modern music production, it is common to use Software Instrument tracks for MIDI. Using a Software Instrument track lets you place a virtual instrument into an Instrument slot in the channel strip. The MIDI on the track then triggers the virtual instrument during playback. A Software Instrument track allows you to route both MIDI and audio signals simultaneously.

Regions Versus Files

Each recording you complete on a track in Logic Pro is stored as a separate audio file on disk. Logic uses pointers to these files, called audio regions, to represent the audio on the associated tracks.

Audio regions can represent the entire audio file or only a portion of it. When you edit an audio region, such as by trimming off the start or end of the audio, you modify the range within the parent audio file referenced by the region. This has no effect on the original file on disk.

By working with regions rather than the original files, Logic Pro can perform non-destructive audio editing. This means the original audio files are unaltered by the edits you perform. As a result, you can easily recover "missing" audio after an edit for later use.

Selecting Tools

Logic Pro includes a variety of tools that you can use to edit the media on your tracks. These are available in two separate menus in the Tracks area menu bar: a Left-Click Tools menu and a Command-Click Tools menu.

Figure 5.35 Tool menus in Logic Pro: Left-Click Tools menu (left) and Command-Click Tools menu (right)

Assigning the Left-Click Tool

The primary tool can be assigned by clicking on the Left-Click Tools menu and selecting the desired tool. The mouse pointer will change to show the active tool and the menu will close.

Figure 5.36 Tools available from the Left-Click Tools menu

The tools you will use most often in Logic Pro include the following:

- **Pointer tool**—Use this tool to select, move, copy, resize, and loop regions and other elements. The appearance of this tool may change based on the current function.

- **Pencil tool**—Use this tool to create a region and to edit its length. You can also use the Pencil tool to add and edit MIDI notes on a software instrument track.

- **Eraser tool**—Use this tool to delete regions, events, markers, folders, and other elements.

- **Text tool**—Use this tool to edit the names of regions, events, markers, folders, and other elements.

- **Scissors tool**—Use this tool to split a region or event into two or more elements.

- **Zoom tool**—Use this tool to zoom in, filling the workspace with a target area by clicking and dragging across it. Click again to revert to the previous view.

- **Marquee tool**—Use this tool to select regions or parts of regions for editing and playback by dragging across them.

 You can quickly access the Left-Click Tool menu by pressing the letter T. To revert back to the Pointer tool, you can press T twice.

Assigning the Command-Click Tool

Logic lets you assign a separate tool to use with a **COMMAND-CLICK** mouse action. This allows you to switch between the primary (left-click) tool and a secondary tool of your choosing by simply pressing the Command key at any time.

To assign the secondary tool, click on the Command-Click Tools menu and select the desired tool. The mouse pointer will change to show the secondary tool whenever you press and hold the Command key.

The tools available in the Command-Click Tools menu are the same as those in the Left-Click Tools menu.

Basic Navigation

As a project grows, both in overall length and in track number, it becomes increasingly important to be able to navigate through the project quickly. In this section, we cover basic navigation techniques for use with playback, recording, and editing in a project.

Using the Transport Controls

The Transport window provides buttons for controlling playback and cursor position. The transport buttons are displayed in the main window's control bar, or in a floating transport window that can be displayed by selecting **WINDOW > OPEN TRANSPORT FLOAT**.

Figure 5.37 Transport buttons in the control bar and floating transport window

The transport buttons provide the following functions (left to right):

- **Rewind**—The Rewind button moves the cursor backwards while held. This function is available both when the transport is stopped and while it is in motion.

- **Fast Forward**—The Fast Forward button moves the cursor forward while held. This function is available both when the transport is stopped and while it is in motion.

- **Play**—The Play button initiates playback from the current cursor location. The play function can also be initiated by pressing the **SPACEBAR** when the transport is stopped.

- **Stop**—The Stop button stops a playback or record pass. The stop function can also be initiated by pressing the **SPACEBAR** when the transport is in motion.

- **Go to Beginning**—Whenever the transport is stopped and the cursor is not at the start of the project, the Go to Beginning button is displayed where the Stop button would otherwise be. This button returns the playhead to the start of the timeline. This function can also be initiated by pressing the **RETURN** key.

- **Record**—The Record button starts a record pass on any record-enabled track(s).
- **Cycle**—The Cycle button enables/disables Cycle mode, creating a range for looping playback or performing a loop recording.

 Cycle mode is discussed later in this chapter.

Zooming and Scrolling in the Main Window

Logic Pro X provides multiple ways to zoom your view in and out. As mentioned earlier in this chapter, you can use the Zoom tool to zoom in on an area. Simply click and drag to draw a selection on one or more tracks and release to fill the screen with the lassoed selection. Afterwards click again on the screen to zoom back out.

 You can temporarily activate the Zoom tool by holding CONTROL+OPTION.

You can also zoom in and out using the playhead. Hold **OPTION** and drag the top of the playhead up or down in the lower section of the ruler to zoom the track area. Dragging up will zoom out, and dragging down will zoom in.

Another option is to use the vertical and horizontal zoom controls in the Tracks area menu bar, beneath the control bar area on the right side of the main window.

Figure 5.38 Auto Zoom buttons and zoom sliders in the Tracks area menu bar

You can click the **VERTICAL AUTO ZOOM** button to automatically adjust track height to fill your workspace with all of the tracks in your project. Similarly, you can click the **HORIZONTAL AUTO ZOOM** button to fill your project's workspace with the longest track, showing you all the regions for tracks displayed in your project. You can also use the vertical and horizontal zoom sliders to manually adjust the zoom for all tracks.

Alternate Zoom Functions

Logic Pro X provides additional zoom keyboard shortcuts that can speed up your workflow when you need to quickly zoom in and out.

- **Zoom In/Out (Horizontal)**—To zoom horizontally from the keyboard, press **Command+Left/Right Arrow**. Each key press zooms by one level. You can also hold **Option+Shift** and scroll the mouse wheel up/down in the workspace.

- **Zoom In/Out (Vertical)**—To zoom vertically from the keyboard, press **Command+Up/Down Arrow**. Each key press zooms by one level. You can also hold **Option** and scroll the mouse wheel up/down in the workspace.

- **Toggling Zoom**—If you have a selection on one or more tracks you can press the **Z** key to zoom in and fill your entire workspace with that selection. This zooms in vertically and horizontally (much like lassoing with the zoom tool). Pressing **Z** a second time toggles back to the original zoom settings.

Horizontal Scrolling

When zoomed in on your project, you will commonly need to scroll the main window left or right to access a desired location on screen. The following options can be used to scroll the main window horizontally:

- Drag left and right using the scroll bar at the bottom of the window.

- Press **Return** to jump to the start of the project.

- Move your playhead incrementally left and right by pressing the **<** or **>** keys.

- Press the **Home** or **End** keys to scroll your displayed workspace forward or backward by a page.

You can also swipe left/right with a Magic Mouse, trackpad, or similar device to scroll the workspace earlier or later. If you have a vertical scroll wheel only, you can hold **Shift** while scrolling to move the screen left/right.

Vertical Scrolling

When working on a project with many tracks (or with tracks set to large display sizes), you will commonly need to scroll the Tracks area up or down to access a desired track on screen. The following options can be used to scroll vertically:

- Press the **Page Up** key to scroll up.

- Press the **Page Down** key to scroll down.

- Scroll up/down with a mouse wheel, trackpad, or Magic Mouse.

Controlling Playback Behavior

You can position the playhead for playback using Transport controls, such as fast-forward and rewind. However, you can also manually position the playhead anywhere within your project to set a playback location. You can also set the scrolling behavior that the Tracks area uses during playback.

Setting a Playback Location

To set the playback location, do one of the following:

- Click anywhere on the lower half of the ruler to position the playhead.

- Double-click in the LCD, enter a time location, and press **RETURN** or **ENTER**. The playhead will move to that location. Alternatively, you can click and drag the bars/beats or time position to move the playhead gradually.

- Click at the desired location on a track with the Marquee tool.

- Click and drag with the Marquee tool on a track or region to make a selection. The selected area will appear highlighted with a dark overlay. When you press Play on the transport (or press the **SPACEBAR**), Logic Pro will play back your selected area.

When you click the **PLAY** button in the Transport (or press the **SPACEBAR**), the project will begin playback from the current playhead location or the start of the selection. Playback will continue until you press the **STOP** button (or press the **SPACEBAR** a second time) or until the playhead reaches the end of your selection.

Setting Scrolling Options

Logic Pro software provides a couple scrolling options that let you specify how the contents of the workspace are displayed during playback and recording: Catch mode and Scroll in Play.

With Catch mode active, the visible section of the project will follow the playhead during playback and recording. You can toggle Catch mode on and off by clicking the **CATCH** button in the main window or by pressing the backtick (`) key.

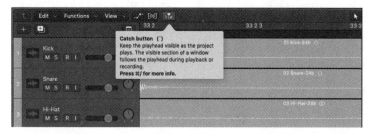

Figure 5.39 The Catch button displayed in the Tracks area menu bar in Logic Pro's main window

By default, Catch mode is automatically engaged when you roll the transport or move the playhead. You can change this behavior by choosing **Preferences > General > Catch** and turning off one or both of these options:

- **Catch When Starting Playback**—With this setting enabled, as soon as you start or pause playback, Catch mode will be enabled.
- **Catch When Moving Playhead**—With this setting enabled, Catch mode will be enabled whenever you position the playhead.

Catch mode can also be used in combination with the Scroll in Play setting. This lets you center the playhead and have it remain stationary as the workspace moves behind it during playback.

- **Scroll in Play**—You can toggle this setting from the View menu in the Tracks area menu bar. Select **View > Scroll in Play** or press **Control+`**.

Figure 5.40 The Edit, Functions, and View menus in the Tracks area menu bar; Catch button shown enabled

 Note that Catch mode must also be enabled for Scroll in Play to function.

Monitoring Your Playhead Location

When navigating a project to set locations for playback, recording, or editing, you'll find it useful to reference the playhead location indicators that Logic Pro's LCD provides for you. You can also use the ruler displays as visual references for bar and beat or minute and second locations.

Location Display in the LCD

You can refer to the LCD to determine the current playhead location. You can click and drag on the bars and beats or time location to move the playhead forward or backward. You can also double-click in the LCD and manually enter a location. Typing a value in the LCD and pressing **Return** will move the cursor to the specified location.

Ruler Displays

In Logic Pro X, the primary ruler displays Musical time in increments of bars and beats by default. This can be useful for music production, allowing you to easily identify where Bar 8 ends and Bar 9 begins, for example.

For some projects, it can be helpful to display absolute time (minutes/seconds/timecode) instead. Doing so will change the ruler increments and the LCD to show time in minutes and seconds, relative to the project start.

To switch the primary ruler to absolute time, select **FILE > PROJECT SETTINGS > GENERAL** and deselect the Use Musical Grid checkbox.

> ⓘ Setting the primary ruler to absolute time can be useful for spoken word recordings or other non-musical projects. In all other cases, it is best to use the secondary ruler for absolute time reference instead.

To display absolute time as a secondary ruler, you can click on the View menu in the Tracks area menu bar, and select **SECONDARY RULER**. Alternatively, you can use the shortcut **COMMAND+OPTION+CONTROL+R**.

> ⓘ You can click on the Display Mode Pop-up menu (down arrow on the right side of the LCD) to display minutes and seconds in your LCD. This will not change the primary ruler.

> ⓘ You can also select **OPEN GIANT BEATS DISPLAY** or **OPEN GIANT TIME DISPLAY** from the Display Mode Pop-up menu to open a floating window showing the current playhead location.

Figure 5.41 A floating Giant Beats Display in Logic Pro

Selections

Making proper selections is an important part of any work you will do in a project. Here, we discuss some of the basics of making and working with selections. Typically, while making selections you'll be using the Marquee tool or the marquee stripe, but you can also use Cycle mode to loop playback within the cycle area. When specifying a selection for recording, you can either use the cycle area or create an auto-punch range.

Using the Marquee Tool and Marquee Stripe

You can click and drag in the tracks area using the Marquee tool to select a portion of the project on one or more tracks. The Marquee tool is set as the Command-Click Tool by default, but you can also select it for your left-click tool. The Marquee tool icon looks like a crosshair.

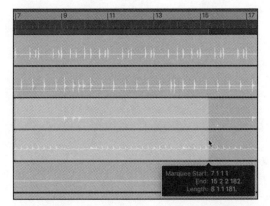

Figure 5.42 A selection created with the Marquee tool in Logic Pro's workspace

The marquee stripe is a thin area at the bottom of the ruler used to create a selection across all tracks. You can make selections in the marquee stripe very much like selecting with the regular marquee tool; however, you may need to enable the marquee stripe to use it.

To make a marquee stripe selection:

1. Choose **VIEW > MARQUEE RULER** from the Tracks area menu bar.

   Figure 5.43 Enabling the marquee stripe

2. Drag across the marquee stripe area of the ruler.

With either method, once you've created a selection, you can adjust the start and end points of the selection. You can hold **SHIFT** (or **COMMAND+SHIFT** if using the Command-click tool), and click/drag to extend the range from either end. You can also use the **LEFT/RIGHT ARROWS** to adjust the selection end point, and **SHIFT+LEFT/RIGHT ARROWS** to adjust the selection start point.

When you have a selection and you engage playback, Logic Pro will automatically move the playhead to the start of your selection. Playback will automatically stop when the playhead reaches the end of the selection.

Creating a Cycle Area

You can cause playback to loop across a specified range in your project by creating a cycle area. When enabled, the cycle area will be displayed as a yellow strip in the ruler. (See Figure 5.44.)

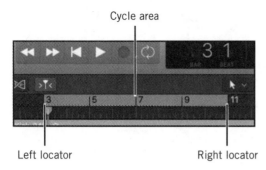

Figure 5.44 The cycle area in the Logic Pro ruler

To set a cycle area, you can click and drag in the upper portion of the ruler. To adjust the cycle length, you can either **SHIFT-CLICK** in the ruler, or drag the left/right locators to adjust their positions. You can also move the entire cycle area, while maintaining its length, by clicking in the center and dragging left or right.

When Cycle mode is active, the playhead will jump to the beginning of the cycle area as soon as you start playback/recording. To enable Cycle mode, click the **CYCLE** button in the control bar or press the **C** key on the computer keyboard. If you have an existing selection made with other methods, you can quickly create a cycle area from the selection by pressing **COMMAND+U**.

Selection Scenarios

To help illustrate some different types of selections, consider the following scenarios:

- **Making a Selection for Playback**—To specify a selection for playback, you can use the marquee tool, the marquee stripe, or a cycle area. When playing back a selected range, all tracks will be included, subject to any Solo or Mute setting you may have made.

- **Making a Selection for Editing**—To select a range for editing or recording, you must make a marquee selection on the associated track or tracks. For example, to delete a portion of the audio on a track, you would select the target area on the track with the Marquee tool and press the **DELETE** key.

- **Making a Selection for Recording**—To specify a recording range, you can use a cycle area or set up an Autopunch range. Autopunch automatically starts and stops recording when you reach specified Punch In and Punch Out locators.

 Recording techniques and autopunch are discussed in detail in Chapter 6.

Review/Discussion Questions

1. What track types are supported in Logic Pro X? (See "Selecting a Track Type" starting on page 103.)

2. Which area of the main window displays the audio waveforms and MIDI performances in a timeline display on tracks in your Logic Pro X project? (See "Main Window" beginning on page 105.)

3. What are some controls available in the Transport? (See "Transport Controls" beginning on page 110.)

4. What is the LCD used for in Logic Pro X? (See "The LCD" beginning on page 110.)

5. What is the primary ruler used for? What measurement units are available for the primary ruler in Logic Pro X? (See "Rulers" beginning on page 114.)

6. How can you access the Mixer window? What are some track controls available in this window? (See "Mixer Display and Window" beginning on page 115.)

7. What is the difference between an External MIDI track and a Software Instrument track? Which type can be used to record MIDI information? Which type can be used to host a virtual instrument? (See "External MIDI Versus Software Instrument Tracks" beginning on page 118.)

8. What are regions in Logic Pro X? How is an audio region different from an audio file? (See "Regions Versus Files" beginning on page 119.)

9. What two menus are available for selecting tools in Logic Pro X? What are some tools available in these menus? (See "Selecting Tools" beginning on page 119.)

10. What is the Cycle button used for in the Transport controls? (See "Using the Transport Controls" beginning on page 121.)

11. What are some ways to zoom in and out in Logic Pro X? What are some ways to scroll the playhead left or right in the main window's displayed workspace? (See "Zooming and Scrolling in the Main Window" beginning on page 122).

12. Which left-click mouse tool can be used to set a playback location by clicking and dragging on a track? (See "Setting a Playback Location" beginning on page 124.)

13. What is Catch mode in Logic Pro X? How does Catch mode relate to the Scroll in Play function? (See "Setting Scrolling Options" beginning on page 124.)

14. What are some ways you can determine the current playhead location? (See "Monitoring Your Playhead Location" beginning on page 125.)

To review additional material from this chapter and prepare for certification, see the Logic Audio Production Basics Study Guide module available through the Elements|ED online learning platform at ElementsED.com.

Exercise 5

Configuring and Working on a Project

🎧 Activity

In this exercise, you will open and configure the display for a pre-configured project. You will then practice making selections and changing the LCD and Timescale Ruler to measure selections in different timebase units.

🕐 Duration

This exercise should take approximately 10 minutes to complete.

◈ Goals/Targets

- Open an existing Logic Pro X project
- Become familiar with basic controls found in the main window
- Toggle the primary ruler between displaying Musical and Absolute values
- Make selections with the Marquee tool

Exercise Media

This exercise uses media files from the song, "Overboard," provided courtesy of The Pinder Brothers.

Written by: Matt Pinder; Performed by: The Pinder Brothers;
Produced by: Scott Reams and The Pinder Brothers*

The media provided for this course may be used for educational purposes only. No rights are granted to use the media for any other personal, commercial, or non-commercial purposes.

* The mix, processing, and media files have been adapted for use in the exercises contained herein.

Getting Started

To get started, you will open an alternative version of the multitrack project you worked on in Exercise 2.

Open the OverboardTRX_Ex5 project:

1. Launch Logic Pro X, if it isn't already running.
2. If the Project Chooser is not visible, select **FILE > NEW FROM TEMPLATE** to display it.
3. If necessary, click the **OPEN AN EXISTING PROJECT** action on the left side of the Project Chooser.
4. Navigate to your Documents folder (or other location where you saved the Logic APB Media Files folder in Exercise 1).
5. Open the 05. Overboard folder within the Logic APB Media Files folder.
6. Select OverboardTRX_Ex5 project from the list of available projects.
7. Click **OPEN**. The pre-made project will open.
8. Choose **FILE > SAVE AS** and name the project Overboard02-*xxx*, where *xxx* is your initials.

Configuring the View

Before starting work on a project, it can be helpful to adjust the view in Logic Pro X to optimize the window displays. In this section of the exercise, you will adjust track heights and close open tabs that obstruct your workspace in the main window.

Close open tabs from the control bar and adjust track heights:

1. Find the control bar toggle buttons for the Inspector, Mixer, and Notepads tabs. Close each of these to free up screen real estate.

Configuring and Working on a Project **133**

Figure 5.45 A crowded Logic Pro workspace with too many tabs open from the control bar

2. Select the Vertical and Horizontal Auto Zoom buttons. (See Figure 5.46.)

Figure 5.46 Automatic Vertical and Horizontal Zoom buttons (left) and zoom sliders (right)

3. Adjust zoom settings using another method of your choosing:

   - Drag the horizontal or vertical zoom sliders.
   - Hold **OPTION** and **OPTION+SHIFT** while scrolling with the mouse wheel (vertical and horizontal zoom, respectively).
   - Press **COMMAND+LEFT/RIGHT ARROWS** (horizontal zoom) and **COMMAND+UP/DOWN ARROWS** (vertical zoom).

   Make sure you're zoomed in enough to see the bar numbers horizontally and waveform data vertically.

Making Selections

The tracks in this project represent a segment of the song "Overboard" by The Pinder Brothers. To get familiar with the audio in this project, you will start by selecting audio segments and setting a cycle area for your selection.

Select audio for the guitar solo on the Lead Guitar track:

1. Activate the **MARQUEE** tool for the Command-click mouse tool.

   Figure 5.47 Marquee tool active in the Command-click menu

2. Hold **COMMAND** to activate the **MARQUEE** tool, then click and drag from Bar 19 to Bar 38 in the Lead Guitar-24b region to select the guitar solo.

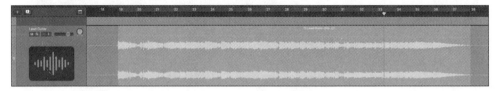

   Figure 5.48 The guitar solo selected in the Lead Guitar-24 region Logic Pro's main window

3. Engage playback. Notice that playback begins at the start of the selection and continues until the playhead gets to the end of the selection at Bar 38, where playback stops.

4. Activate the Cycle button by clicking it in the transport or by pressing **C**.

5. Next press **COMMAND+U** to match the cycle area to the selection. You can now listen to the project play back the guitar solo section on a loop.

   Figure 5.49 The cycle area set over the duration of the guitar solo

Next, you will change the LCD settings to show both musical and absolute time formats. You will also display the secondary ruler to measure your selection in minutes and seconds.

Change the LCD settings and display the secondary ruler:

1. Click on the triangle on the right side of the LCD, next to the project information section.

2. Choose **BEATS & TIME** from the pop-up menu. (See Figure 5.50.) The LCD will update to show absolute time along with bar and beat measurements.

Figure 5.50 The LCD set to Beats & Time display

3. Click on the **View** menu in the Tracks area menu bar and select **Secondary Ruler**. The secondary ruler will display, showing absolute time in minutes and seconds.

 Press Command+Option+Control+R to display the secondary ruler from the keyboard.

4. Click on the Display Mode pop-up selector next to the LCD display a second time and select the Beats & Project (Large) display.

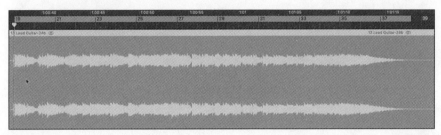

Figure 5.51 The guitar solo selection displayed with the secondary ruler showing absolute time

5. Use the indicators in the LCD and the secondary ruler to determine the start and end locations for the cycle area in minutes and seconds.

Finishing Up

Feel free to experiment further with the project. Practice making selections with the marquee tool, setting new cycle areas, and zooming using different methods. When finished, return to the start of the project and save the work you've done.

Saving the project changes:

1. Click the **Go to Beginning** button in the Transport controls to return the playhead to the start of the timeline.

2. Choose **File > Save** to save the project.

3. Choose **File > Close Project** to close out of the project.

That completes this exercise.

Chapter 6

Logic Pro X Concepts, Part 2

...*What You Need to Know to Work with Logic Pro X*...

We begin this chapter with an overview of recording audio and MIDI performances using Logic Pro X. Next, we explore options for importing existing audio and MIDI data into your project. And finally, we take a look at basic editing techniques that can be used to fine-tune audio and MIDI performances.

◎ Learning Targets for This Chapter

- Learn how to record audio
- Learn basic MIDI recording techniques
- Learn how to import audio and MIDI
- Understand how the Snap to Grid setting affects selections and edit operations
- Learn basic techniques for working with regions and selections

Key topics from this chapter are illustrated in the Logic Audio Production Basics Study Guide module available through the Elements|ED online learning platform. Sign up at ElementsED.com.

When starting a new Logic Pro X project, the first step is typically to either record or import some audio or MIDI data into your project. This can be accomplished in a number of ways, but generally you'll either record new regions directly onto an Audio or Software Instrument track in your project or import existing files from various locations on your computer.

Setting Up for Recording

When beginning a music project, it can be beneficial to set the project tempo for Logic Pro X prior to starting to record. This will allow you to use a metronome click as a tempo reference during recording, helping the musicians keep the performance consistently in time. Additionally, recording to a click will ensure that the music aligns to the bars and beats of the primary ruler. This means that any selections and edits you make can easily be aligned to the music using the Snap to Grid feature.

 The project tempo defaults to 120 beats per minute (BPM) in Logic Pro X.

To set the project tempo:

1. Click the triangle on the right side of the LCD and choose **BEATS & PROJECT** from the pop-up menu. The LCD area will expand to show the Tempo field and other project information.

2. Do one of the following:

   - Double-click in the Tempo field to highlight the default value of **120**. Type in the tempo value that you'd like to use (in BPM) and press **RETURN** to commit the value.

   Figure 6.1 Tempo field selected in the LCD Display

   - Click and hold the Tempo field and drag up or down to scroll to the tempo value that you'd like to use.

To turn on the metronome, do one of the following:

- Click the **METRONOME** button in the control bar.
- Press **K**, the keyboard shortcut assigned to toggle the Metronome on and off.

Recording Audio

Recording audio is one of the most fundamental aspects of working with any modern DAW. Working in Logic Pro X gives you non-linear access to all of the audio in your project. This means you can jump directly to any location without needing to fast–forward or rewind like you would with a tape recorder.

Logic Pro X provides a truly intuitive workflow for quickly getting your musical ideas recorded onto a track. And thanks to the huge storage drives included with most modern computers, you'll rarely have to worry about running out of recording space.

Making a Record Selection

Before you begin recording, you'll need to specify where you want the record pass to begin and end. Audio engineers call the beginning and end of the target area the "punch-in point" and "punch-out point," respectively. In Logic Pro X, you can specify a range using Cycle mode or the autopunch settings. You can also choose to specify only the starting point and opt out of setting an end point, letting Logic Pro X continue recording until you manually stop the record pass.

Figure 6.2 An autopunch selection prepared for a lead vocal track

To set a record location without specifying a stopping point:

- Move the playhead to the position on the ruler where you want to begin recording.

To set a record range using Cycle:

1. Click the **CYCLE** button in the control bar (or press **C**) to activate the cycle area.

2. Move the pointer over the left and right edges of the cycle area and drag to the desired start and end locator positions on the ruler. You can also move the entire cycle area by grabbing from the middle and dragging to the left or right.

To set a record range using autopunch:

1. Display the Autopunch button in the control bar by **CONTROL-CLICKING** the control bar and selecting **CUSTOMIZE CONTROL BAR AND DISPLAY**. Then select the check box for **AUTOPUNCH**. Alternatively, you can **OPTION+COMMAND-CLICK** anywhere in the ruler.

2. Turn on autopunch by clicking the **AUTOPUNCH** button in the control bar so that it becomes active.

Figure 6.3 The Autopunch button in the control bar

3. Move the pointer over the left and right edges of the autopunch area (red strip in the ruler) and drag to the desired punch in and punch out locator positions. You can also move the entire autopunch area by grabbing from the middle and dragging to the left or right.

Record Enabling Audio Tracks

Once you've set a location for recording, you'll need to specify which tracks to record on. Logic Pro X conveniently record enables any track you've selected. If the Record Enable button is visible on a selected track, the letter **R** will be red, indicating it is automatically enabled to record.

Figure 6.4 A record enabled audio track in the Main window

 To record multiple tracks in Logic Pro X simultaneously, click the RECORD ENABLE button for each additional track you wish to record to.

Monitoring Record Enabled Audio Tracks

Whenever you click **PLAY** (or press the **SPACEBAR**), you will hear any regions already in place on your tracks. Tracks that have no regions will remain silent. This is true even when a track is record-enabled and you begin playback: you will hear any existing material on that track and *not* the live input from your audio interface.

This can be a little confusing, especially if you are trying to audition a part before starting to record. Fortunately, you can listen to (or *monitor*) the live input at any time by enabling the target track's Input Monitoring control (the **I** button).

Figure 6.5 A track with the Input Monitoring button enabled in the Main window

To enable Input Monitoring on an Audio track in Logic Pro X:

1. Locate the desired track in the Inspector, Mixer, or Main window.

2. Do one of the following:

   - Click the **INPUT MONITORING (I)** button on the track.
   - Select the track and press **CONTROL+I** on your computer keyboard.

Initiating a Record Take

Once you have record enabled one or more tracks, you can initiate a record take in Logic Pro X. The most common technique for beginners is to press the **RECORD** button in the Main window control bar.

— Record button

Figure 6.6 The Transport controls in the Main window control bar

Logic Pro X also provides a number of keyboard shortcuts that can be used to initiate record. If you record frequently, these are some of the first shortcuts that you'll want to learn!

To immediately begin recording, do one of the following:

- Click the **RECORD** button in the Transport controls.
- Press the **R** key.
- Press the **[*]** key on the numeric keypad.

Stopping a Record Take

If you're using autopunch, Logic Pro X will automatically stop recording when it reaches the punch out locator. However, if you are recording without autopunch, the record take will continue until you manually stop the transport. Logic Pro X provides several techniques for stopping a record take.

To immediately stop a record take, do one of the following:

- Press the **STOP** button in the Transport controls.
- Press the **SPACEBAR**.
- Press the **[0]** key on the numeric keypad.

At times you may want to stop a take prematurely due to a problem in the record pass. Often a musician will make a mistake right at the beginning of the take and won't want to continue. Fortunately, Logic Pro X provides a helpful shortcut that will both stop the take and delete the partially recorded region. Obviously, you'll only want to use this option if you're absolutely certain that the take is not worth keeping.

To immediately stop a record take and delete the recorded region:

- Press **COMMAND+PERIOD**. Logic Pro X will abort the pass, discard the recorded audio without saving it to disk, and return the playhead back to the last play position.

Recording Additional Takes

When recording, sometimes you will want to record multiple versions (or *takes*) of a section. For example, you may be improvising or may need to perform a part a few times to capture the best performance. In Logic Pro X, when you overlap recordings on the same track, a Take Folder is created containing each recording pass. After recording, you can choose the best take or edit parts of different takes into a new version (or *comp*).

Figure 6.7 Take Folder created while recording multiple takes

By default, Logic Pro X will create a Take Folder automatically after any overlapped recording.

 To change these settings, select RECORD > OVERLAPPING AUDIO RECORDINGS > CYCLE and select the desired option.

Logic Pro X gives you the option to record takes with cycle mode on or off. When cycle is on, recording begins at the left locator of the cycle area. After each cycle pass, a take is created and the playhead seamlessly starts another pass from the left locator. When cycle is off, you must manually start and stop recording as well as position the playhead where you want to start each take.

To record multiple takes with cycle on, do the following:

1. Enable Cycle mode and set a cycle area.
2. Click **RECORD** (or press **R**).

3. Record as many takes as you like. A take folder will automatically be created. Each cycle pass will be added to the take folder.

 When cycle mode is on, the playhead will automatically start from the left locator position of the cycle area when playing or recording.

To record multiple takes with Cycle mode off, do the following:

1. Position the playhead where you want to start a recording.
2. Click **Record** (or press **R**) to start a record pass, and then click **Stop** (or press **Spacebar**) when finished.
3. Reposition the playhead where you want to start a second recording.
4. Repeat the recording process for as many additional takes as you like.

Recording MIDI

Although the MIDI communication protocol is more than 30 years old, understanding how to record and edit MIDI data is more relevant today than ever. Advances in virtual instrument design and computer processing speed over the past decade have made virtual instruments an indispensable part of the modern music production workflow. The fundamental workflow for recording MIDI is similar to that for recording audio. However, some important differences apply in MIDI production. Fortunately, you don't need to master the entire field of modern synthesis to start making music with the MIDI feature set in Logic Pro X!

Monitoring a MIDI Controller

The monitoring workflow is one area that differs significantly when recording MIDI versus recording audio. This is because an audio signal can be monitored directly, whereas MIDI data must be routed to an instrument to become an audible signal. The routing can seem a little complicated at first, but once you understand the basics you'll be able to take advantage of this powerful technology.

 For basic information on routing MIDI data to virtual instruments, see "Tracking with Virtual Instruments" in Chapter 4.

Routing MIDI data to a virtual instrument in Logic Pro X is fairly simple. After creating a software instrument track, you can select the instrument you want to use by clicking on the Instrument plug-in slot in the Inspector or by selecting a patch from the Library.

Figure 6.8 Clicking the plug-in slot in the Inspector

Figure 6.9 Selecting the Liverpool Bass patch from the Library (displayed using the Library button on the control bar)

Record Enabling Software Instrument Tracks

As with Audio tracks, Logic Pro X automatically record enables any selected software instrument track. If the Record Enable button is visible on a selected track, the letter **R** will be red indicating it is automatically enabled to record.

Figure 6.10 A record enabled software instrument track in the Main window

Overlap Recording on Software Instrument Tracks

When recording over an existing MIDI region, you can create a take folder in the same way you do for audio. Additionally, since MIDI information is handled differently than audio, you have the option to merge multiple takes into one region. One benefit of this feature is that it allows you to divide a performance into multiple layers making it easier to record more difficult passages. For example, you can record individual parts of a complex drum beat one at a time.

By default, Logic Pro X will merge overlapped takes when recording MIDI. Just like with audio tracks, Logic Pro X gives you the option to record overlapping takes with Cycle mode on or off.

To merge multiple takes into one region on a selected software instrument track, do the following:

1. Select **Record > Overlapping MIDI Recordings > Cycle > Merge**. (Select this option for both **Cycle On** and **Cycle Off**.)

2. Press **Record** (or press **R**).

3. Record as many takes as you like. Each recording pass will merge with the existing region.

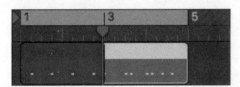

Figure 6.11 Recording multiple takes to merge into a single region.

 MIDI performances are captured by storing NOTE ON and NOTE OFF information; be careful when merging duplicate notes over the same spot in time or rhythm. Logic Pro X will keep each instance, which may cause undesirable results.

To create a take folder on a selected software instrument track, do the following:

1. Select **Record > Overlapping MIDI Recordings > Cycle > Create Take Folder**. (Select this option for both **Cycle On** and **Cycle Off**.)

2. Press **Record** (or press **R**).

3. Record as many takes as you like. A take folder will automatically be created, and each recording pass will be added to the take folder.

Importing Audio and MIDI

If you're not ready to record in Logic Pro X, or if you'd like to use existing media from your hard drive instead, you can import audio or MIDI files into your Logic Pro X project to get started. There are a number of ways to import audio and MIDI files into Logic Pro X.

Supported Audio Files

Before you attempt to import an audio file, you'll want to make sure that it is in a format that can be imported into Logic Pro X. The following file types can be imported:

- AAC Audio
- CAF
- AIFF
- WAV
- MP3
- QuickTime (Mac only)
- ReCycle (RCSO, RCY, REX, and REX2)
- Sound Designer I and II (SDII)

Importing from the Desktop

An easy way to get started importing is to use the drag-and-drop method. Logic Pro X fully supports drag-and-drop importing of audio and MIDI files from the Finder. Simply select the desired files on your computer and drag them into your Logic Pro X project.

Importing from Media Browser and All Files Browser

Another way to import audio and MIDI files into Logic Pro X is to use the built-in Media Browser or All File Browser. These browsers work similar to a Finder window. Use the Media Browser to browse files within other Logic projects, GarageBand projects, and your iTunes library. Use the All Files Browser to browse files on any mounted volume on your computer. In both browsers, you can manually browse or use the search field.

To import from the Media Browser, do the following:

1. Click the **BROWSER** button in the control bar, then select the **MEDIA** tab.
2. Select **AUDIO** or **MOVIES** to navigate to the type of content you want, and select the location or folder containing files you wish to import.

3. Drag-and-drop the desired files into the Tracks area.

To import from the All Files Browser, do the following:

1. Click the **BROWSER** button in the control bar, then select the **ALL FILES** tab (or press **F**).

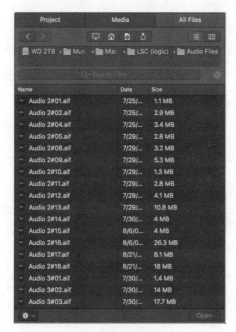

Figure 6.12 The All Files browser

2. Navigate to the location or folder containing files you wish to import.

3. Drag-and-drop the desired files into the Tracks area.

Using the Import Command

Audio and MIDI files can also be imported into Logic Pro X using the **FILE > IMPORT** command. This option uses a basic file dialog box to let you locate files anywhere on your computer's hard drive. Imported files will be placed on the selected track, at the Playhead position in Logic Pro X.

To import audio, do the following:

1. Select an audio track in the Tracks area.

2. Choose **FILE > IMPORT > AUDIO FILE**.

3. Select the audio file you wish to import.

4. Click the **IMPORT** button to add the file to the selected track.

To import MIDI, do the following:

1. Select a software instrument track in the Tracks area. (If no track is selected, a new track will be created with a default software instrument.)

2. Choose **File > Import > MIDI File**.

3. Select the MIDI file you wish to import.

4. Click the **Import** button to add the file to the selected track.

   You will be prompted with a dialogue box asking whether you would like to import tempo information to your project.

   - Select **No** to have the imported MIDI follow the current project tempo.
   - Select **Import Tempo** to have the project adopt the tempo of the MIDI file.

Figure 6.13 The Import tempo information dialog box

Using Apple Loops

Logic Pro X comes with 10,000 Apple Loops including drum beats, rhythm sections, phrases, leads, and much more that can be added to any project. Whether for a professional production or drafting an idea, Apple Loops can be used as is or edited to suit your needs without limitations.

Three types of Apple Loops are available: Audio Apple Loops, Software Instrument Apple Loops, and Drummer Apple Loops. Audio and Software Instrument Loops function exactly as discussed earlier in this chapter. However, Drummer Loops work a little different. Drummer Loops are like software instruments that use a special interface to customize the performance.

 Drummer Loops are beyond the scope of this course. For additional information, please see the Logic Pro X User Guide.

To start using Apple Loops in your project, do the following:

1. Click the **Loop Browser** button in the Control bar (or press **O**).

2. Search for loops by selecting an instrument category or genre. You also have the option to use the search field and use keywords or titles to locate candidate files for your project.

3. Drag a selected file onto the Track area to add it to your project.

> ⓘ Apple Loops will automatically adapt to the project tempo when imported. Other audio files won't always do this. The audio files included as Apple Loops are configured to follow the project tempo and align to the grid to stay on beat.

Working in the Tracks Area

Before you start editing regions, it helps to have a good understanding of how editing and aligning will work in the Tracks area in Logic Pro X. This will impact the movement and placement of regions and the function of various tools. You should also know how to modify the grid value so that regions will align properly when working with Snap.

Snap to Grid

The Tracks area shows a grid to help with aligning audio and MIDI regions in your project. Any time you move or edit a region or parameter, the Snap to Grid setting determines the behavior of these actions.

By default, Snap to Grid is turned on in Logic Pro X. In most editing and arranging situations, this is preferred because it keeps musical elements aligned relative to the bars and beats of the project. When moving or editing regions in this mode, actions will snap to the defined grid.

Figure 6.14 The blue power button next to the Snap pop-up menu indicates Snap to Grid is active.

Setting the Snap Value

When Snap to Grid is active, you'll want to set the Snap value to an appropriate size. You can set the Snap value using the Snap pop-up menu in the Tracks area menu bar.

To set the Snap value:

- Click the Snap pop-up menu and select the desired grid size or snap behavior.

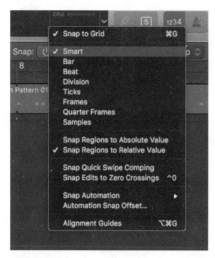

Figure 6.15 The Snap pop-up menu in the Tracks area menu bar

By default, the Snap value is set to Smart. With this setting, movements and edit actions will snap to the nearest bar, beat, division, etc., depending on the current ruler division value and zoom level. As the name implies, the "smart" choice of a snap value is made dynamically depending on your conditions.

> When dragging or trimming a region with Snap to Grid active, you can press and hold CONTROL+ SHIFT to override the snap function. Depending on zoom level, the grid will switch to ticks or samples.

You have the option to select a specific, fixed snap value if you prefer. You can select the value to Bars, Beats, Divisions, Ticks, Frames, Quarter Frames, and Samples. Each of these represents successively smaller divisions of time.

Two additional settings are worth noting in the Snap pop-up menu. When adjusting or editing regions, actions can snap to Absolute or Relative values.

- *Snap Regions to Absolute Value:* Edits will snap the region start to the nearest Snap grid division regardless of its original starting position.

- *Snap Regions to Relative Value:* Edits will maintain a region's existing offset from the Snap grid. This is the default behavior in Logic Pro X.

> *Snap Region to Relative Value* is useful when moving a region whose starting point is not on the grid. The relative starting position will be preserved, maintaining the correct alignment with the bars and beats of the project.

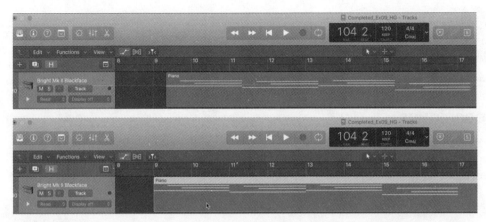

Figure 6.16 Moving a clip with Absolute Snap values (Before: top; After: bottom)

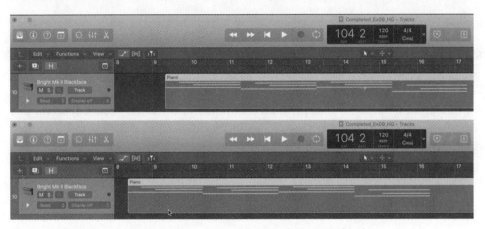

Figure 6.17 Moving a clip with Relative Snap values (Before: top; After: bottom)

Working with Regions and Selections

Once you've recorded or imported some audio or MIDI material, it's time to start editing!

Selecting a Tool for Editing

To get started editing, you'll familiarize yourself with the tools available in Logic Pro X. Most of the time, you will use the Pointer tool. As you move over specific areas, the pointer will change to represent the appropriate tool for the actions you are likely to need to perform. (See Figures 6.18 and 6.19.)

Figure 6.18 The Pointer tool

Figure 6.19 The Pointer tool functioning as the Trim pointer near a region boundary

Basic Editing Techniques

DAWs provide a variety of ways to edit and rearrange audio and MIDI clips. You'll become familiar with a multitude of editing techniques in Logic Pro X as you work, but for now we'll cover the basic techniques you'll need to get started:

- Selecting regions
- Moving regions
- Cutting, Copying, and Pasting regions

These are the basic building blocks that you'll want to learn. Let's go through each in some detail.

Making Selections

Selecting material is one of the most basic and frequently used techniques in the DAW world. It serves as a building block for further editing. Many commands require a region (or portion of a region) to be selected before the command can be executed.

To select an entire region, do the following:

- Using the Pointer tool, click on the target region.

To select a portion of a region:

- Using the Marquee tool, click and drag a selection within the target region.

 The Marquee tool is activated by holding the COMMAND key, by default.

Figure 6.20 A partial selection made using the Marquee tool

Selections with the Marquee tool can include multiple regions, partial regions, and blank areas between regions. Edit operations will utilize the entire selection and preserve blank areas, as appropriate.

Moving Regions

At times you may want to move a region to a new location on a track or onto a completely different track. While you can use Cut/Copy and Paste for this purpose, it's usually faster to simply drag the region to the desired location using the Pointer tool.

To move a region to a new location on the same track:

- Using the Pointer tool, click and drag the region horizontally.

To move a region to a different track:

- Using the Pointer tool, click and drag the region vertically.

 When dragging a region to a different track, you can press SHIFT to constrain the region and prevent it from moving left or right.

Cutting, Copying, and Pasting Regions

For edits involving partial clips or long selections, you'll often want to use Cut/Copy/Paste commands. These commands can be accessed at the top of the Edit menu in Logic Pro X, but you'll definitely want to use the keyboard shortcuts to save time. Logic Pro X uses standard keyboard shortcuts for each of these operations:

- Cut operation—**COMMAND+X**
- Copy operation—**COMMAND+C**
- Paste operation—**COMMAND+V**

Another technique that provides Copy/Paste behavior is the Option-drag method. This operation requires only one key and mouse combination to perform the action of copying and pasting.

To use Option-drag:

- Hold **OPTION** while clicking on a region and dragging it to the desired location. A duplicate copy of the region will be placed at the new location.

Advanced Editing Techniques

Beyond selecting and moving your audio or MIDI data, you'll often need to use slightly more advanced editing techniques. Some useful options include trimming the start/end of a region, splitting a region into two or more separate regions, and adding fades to regions. All of these operations use non-destructive editing in Logic Pro X.

Non-Destructive Audio Editing

Non-destructive editing is an important concept to understand when editing audio data in a DAW. Essentially, non-destructive editing means that any changes you make to an audio region in Logic Pro X will not affect the original audio file stored on disk. This gives you the freedom to try any number of different edits without fear of permanently modifying the original file.

 A few rare operations perform destructive editing in Logic Pro X. For example, In the File Editor, if you draw over an audio region's waveform with the Pencil tool, you will be editing the original audio file on disk.

Editing MIDI Regions

While editing audio data is almost always non-destructive, the same is not true for MIDI data. This is because MIDI regions in Logic Pro X are stored within the project file, and not in a separate file on disk. As a result, some MIDI editing commands *will* modify the original region. Although it is possible to use the undo command to revert a region back to the original, this option is typically useful only immediately after making an unwanted change.

Trimming the Start and End of Regions

Trimming operations are used constantly when editing audio in Logic Pro X. A trim will essentially add or subtract audio from the beginning (head) or end (tail) of a region.

Trimming allows you to make a region shorter by removing audio from the start or end of the region. You can also make a region longer again at *any time* in the future by trimming it back out. In other words, regions still have access to *all* of the original audio data from the source file. This lets you recover any missing audio that you later decide to keep.

 You will only be able to trim the clip start or end up to the boundary of the underlying audio data in the original file. Once you reach the file boundary, you won't be able to trim the audio any further.

To trim off the start of a region:

1. Using the Trim pointer, click and hold inside the region near the lower-left edge.
2. While continuing to hold the mouse, drag to the right to remove audio and move the region start later.

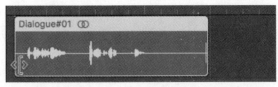 

Figure 6.21 Dialogue clip before trimming (left) and after trimming (right)

To trim off the end of a region:

1. Using the Trim pointer, click and hold inside the region near the lower-right edge.

2. While continuing to hold the mouse, drag to the left to remove audio and move the region end earlier.

Splitting a Region into Two or More Regions

Another very common editing technique involves splitting a region into two or more smaller regions. In Logic Pro X, you can use a simple edit command to perform this function: **EDIT > SPLIT > REGIONS AT PLAYHEAD**.

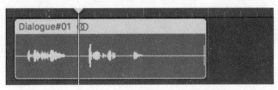 

Figure 6.22 A single region before separating (left) and the result of splitting at the playhead location (right)

To split a region into two regions:

1. Using the Pointer tool, select the target region.

2. Move the playhead to the desired location inside the selected region where you want to make a split.

3. Do one of the following:

    - Select **EDIT > SPLIT > REGIONS AT PLAYHEAD**.

    - Press **COMMAND+T**.

 You can also use the Marquee tool to split a region by placing the pointer over the location you wish to edit and **DOUBLE-CLICK** while holding **COMMAND**. You don't need to move the playhead to perform this action.

Fading Regions

Last but certainly not least, you can apply fades to your audio regions. Fades can be used to gradually fade audio in and out. They are also commonly used to prevent undesirable pops and clicks at the beginning and end of regions. Here, we'll focus on the basics of creating and editing fades in Logic Pro X.

To create a fade in or fade out:

1. Select the Fade tool from the Left-Click Tool menu in the Tracks area menu bar.

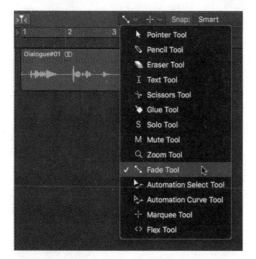

Figure 6.23 Selecting the Fade tool from the Left-Click Tool menu

2. Drag over the start or end point of an audio region to create a fade.

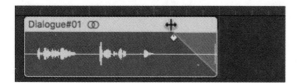

Figure 6.24 A Fade Out created on an audio region

3. Optionally, you can make adjustments to the fade:

   - Adjust the length of the fade by moving the start or end point.
   - Adjust the fade curve by dragging the fade curve (diagonal line) to the left or right.

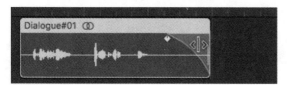

Figure 6.25 Adjusting the Curve of the Fade Out

In addition to fade–ins and fade–outs, the Fade Tool can be used to create crossfades. Crossfades are useful to smooth out an edit location, by fading out the audio from before the edit point while simultaneously fading in the audio from after the edit point.

To create a crossfade:

1. Select the Fade Tool from the Left-Click Tool drop down menu in the Tracks area menu bar.
2. Click and drag across the junction between two adjacent clips. A crossfade graphic will appear.
3. Adjust the fade settings as desired, using the methods described above.

Figure 6.26 Making a Crossfade selection (left) and the resulting Crossfade (right)

If you decide to modify a fades later, you can select the Fade Tool to make adjustments. You can also make fade adjustments from the Region Inspector.

Review/Discussion Questions

1. What process can you use to specify a starting point ("punch in") for recording without specifying a stopping point ("punch out")? (See "Making a Record Selection" beginning on page 139.)

2. What are some ways that you can specify both a starting and stopping point for a record take in Logic Pro X? (See "Making a Record Selection" beginning on page 139.)

3. What are three ways that you can immediately begin a record take in Logic Pro X? (See "Initiating a Record Take" beginning on page 141.)

4. What are three ways that you can immediately stop a record take? How can you immediately stop a record take and delete the recorded region? (See "Stopping a Record Take" beginning on page 141.)

5. What is a Take Folder and how is it created? (See "Recording Additional Takes" beginning on page 142.)

6. What are two ways that you can record multiple takes? Where do you find takes? (See "Recording Additional Takes" beginning on page 142.)

7. How is merging MIDI useful when recording drum parts? (See "Overlap Recording on Software Instrument Tracks" beginning on page 145.)

8. What types of audio files can be imported into Logic Pro X? (See "Supported Audio Files" beginning on page 146.)

9. What are some differences between the Media Browser and the All Files Browser? How would you go about accessing each? (See "Importing from Media Browser and All Files Browser" beginning on page 146.)

10. Where are imported files placed when you use the **FILE > IMPORT** command? (See "Using the Import Command" beginning on page 147.)

11. What three types of Apple Loops are available from the Loop Browser in Logic Pro X? (See "Using Apple Loops" beginning on page 148.)

12. How can you set the Snap value for your project? What are available options for the Snap value? (See "Setting the Snap Value" beginning on page 149.)

13. Which tool can you use to select an entire region by clicking on it? Which tool must be used to select a portion of a region? (See "Making Selections" beginning on page 152.)

14. Which tool must be used to drag and drop a region to a new location in the session? (See "Moving Regions" beginning on page 153.)

15. When is audio editing a destructive process? What about MIDI editing? (See "Advanced Editing Techniques" beginning on page 154.)

16. What Logic Pro X command is used to split a region into two separate regions? (See "Splitting a Region into Two or More Regions" beginning on page 155.)

17. How would you go about adding a fade to a region? What are three kinds of fades you can use in Logic Pro X? (See "Fading Regions" beginning on page 155.)

To review additional material from this chapter and prepare for certification, see the Logic Audio Production Basics Study Guide module available through the Elements|ED online learning platform at ElementsED.com.

Exercise 6

Importing and Editing Regions

🎧 Activity

In this exercise, you will learn how to perform basic editing techniques in Logic Pro X. You'll begin by creating a new project. Then you'll import audio files and a MIDI file into the project and configure the tracks. Finally, you'll do some basic audio editing to remove unnecessary content from an audio region.

⏲ Duration

This exercise should take approximately 15 minutes to complete.

⊕ Goals/Targets

- Create a new blank project document
- Import audio and MIDI into the project
- Assign virtual instrument patches from the Library for the MIDI parts
- Remove audio at the beginning of two drum tracks for a 4-bar intro
- Save your work for use in Exercise 7

Exercise Media

This exercise uses media files from the song, "Lights," provided courtesy of Bay Area band Fotograf.

Written by: Zack Vieira and Eric Kuehnl; Performed by: Fotograf

The media provided for this course may be used for educational purposes only. No rights are granted to use the media for any other personal, commercial, or non-commercial purposes.

Getting Started

To get started, you will create a new project to use for the exercise.

Create the project:

1. Create a new project by selecting **FILE > NEW FROM TEMPLATE** or pressing **COMMAND+N**.

2. When the Logic Pro X Project Chooser appears, select **EMPTY PROJECT**.

3. Click the **CHOOSE** button at the bottom of the Project Chooser to create the project. A new project will be created and will display on screen.

4. Logic Pro X will prompt you to create a new track before continuing. Select **AUDIO** and click **CREATE**. The project will open in the main window.

5. Save the project by selecting **FILE > SAVE** or pressing **COMMAND+S**. A Save dialog box will display.

6. Name the project **Lights-xxx**, where *xxx* is your initials.

7. Under **ORGANIZE MY PROJECT AS A:** choose the Folder radio button. Under **COPY THE FOLLOWING FILES INTO YOUR PROJECT:** check the box for Audio Files.

Figure 6.27 The Save dialog box with the correct settings to save the project

8. Click **Save** at the bottom of the dialog box.

9. Maximize or resize the main window as needed for a full-screen view.

Importing Audio

In Exercise 2, you imported audio files by dragging and dropping into your first project. Here, you'll use the same process to start a second project. You'll be using this new project for the remainder of the exercises in this book.

Import audio for your project:

1. Open the Browser by clicking the **Browser** button in the control bar or pressing the **F** key.

2. Select the **All Files** tab, and navigate to your Documents folder (or other location where you saved the Logic APB Media Files folder in Exercise 1).

3. Open the 06. Lights folder within the Logic APB Media Files folder.

4. Select all the audio files and drag them onto the first measure in the Tracks area of the main window.

5. When the **Add Selected Files to Tracks** dialog box displays, select the **Use existing tracks** radio button.

6. Click **OK** to complete the import. Nine new tracks will be added to your project, and the imported audio regions will be placed on your ten tracks (the original audio track plus the nine new tracks).

Figure 6.28 An overview of the project with the imported audio tracks

Importing MIDI

In this part of the exercise, you will import two MIDI clips using a Standard MIDI file.

Import the MIDI data:

1. Choose **FILE > IMPORT > MIDI FILE** to display the **OPEN** dialog box for your system.

2. In the dialog box, locate and select the Exercise-06.mid file inside the 06. Lights folder.

3. Click **IMPORT** to import the file. You will be prompted with a dialog box asking whether you would like to import tempo information to your project.

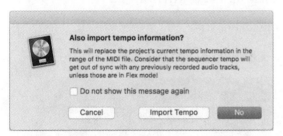

Figure 6.29 The import tempo information dialog box

4. Click the **NO** button to proceed without importing tempo. You might also be prompted with another dialog box asking what SMPTE frame rate to use. In most situations, and for this exercise, it's a good idea to use the recognized value. If you get this message, select **USE 29.97**.

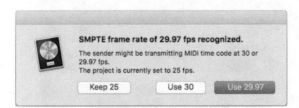

Figure 6.30 The SMPTE frame rate dialog box

Two new Software Instrument tracks will be added to your project.

5. Select the two MIDI regions and move them so that the Piano MIDI region starts at Bar 25 and the Strings MIDI region starts at Bar 77.

Assigning Virtual Instruments

When importing a MIDI file, Logic Pro X reads information in the file that may indicate the original virtual instruments used. Logic Pro uses this information to choose software instrument patches from the library. If no instrument information is specified in the file, Logic Pro X will default to a virtual piano patch.

In this exercise, default piano presets have been selected, so you'll next need to assign appropriate software instruments to the new tracks. You'll use the Library to select a patch for each software instrument track.

To select a software instrument patch, do the following:

1. Select the first software instrument track containing the **Piano** MIDI region.
2. Click the **LIBRARY** button (or press **Y**) to open the Library.
3. Choose the **VINTAGE ELECTRIC PIANO > BRIGHT MK II BLACKFACE** patch.

Figure 6.31 Selecting a patch from the Library

4. Repeat the above process on the second software instrument track containing the **Strings** MIDI file, assigning the **ORCHESTRAL > STRINGS > SMART STRINGS** patch.
5. Close the Library when finished.

Preview the project:

1. Press the **SPACEBAR** to begin playback and listen to the project in its current state.

2. Pay particular attention to the number and type of instruments that you hear. Feel free to solo and mute tracks to isolate certain elements and get familiar with the different parts.

Removing Unwanted Audio

In this part of the exercise, you will remove some drum parts at the beginning of the song. This will create a more dramatic intro featuring the Synth Bass and Electric Piano parts.

Remove the unwanted audio:

1. Use **VERTICAL ZOOM SLIDER** and **HORIZONTAL ZOOM SLIDER** (or other method of your preference) to zoom in on the 01 Kick track for a better view.

2. Set the Snap value to **BAR** using the Snap Value drop-down menu in the Tracks area menu bar.

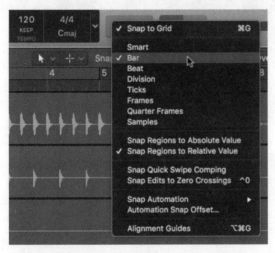

Figure 6.32 Setting the Snap value to one bar

3. Using the Trim pointer, click the bottom-left edge of the 01 Kick region and trim the first four bars off. A help tag will pop up showing position and length details. Trim to Position: 5|1|1|1.

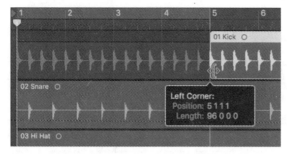

Figure 6.33 Left Corner Position help tag

4. Repeat Step 3 for the region on the 02 Snare track.

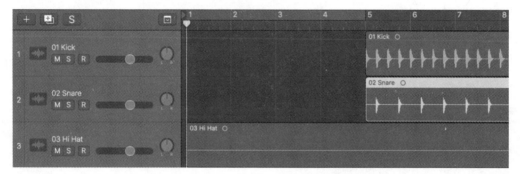

Figure 6.34 The first four bars removed from the 01 Kick and 02 Snare tracks

Finishing Up

To complete this exercise, you will need to save your work and close the project. You will be reusing this project in Exercise 7, so it's important to save the work you've done.

Before you wrap up, you should also listen to the project to hear the changes you've made.

Finish your work:

1. Choose **FILE > SAVE** (or press **COMMAND+S**) to save the changes you've made to the project.

2. Press **RETURN** to position the Playhead at the beginning of the project.

3. Press the **SPACEBAR** to begin playback. Only the synth bass and electric piano should be audible for the first four bars, creating a nice intro.

4. When finished listening to the project, press the **SPACEBAR** a second time to stop playback.

5. Choose **FILE > CLOSE PROJECT** to close the project.

That completes this exercise.

Chapter 7

Mixing Concepts

...*What You Need to Know to Mix a Project*...

This chapter introduces you to mixing concepts, from setting levels and panning to the role of EQ and dynamics processing on the tracks in a mix. We discuss how to keep a mix from clipping, the difference between gain-based processing and time-based processing, and the differences between insert processing and send-and-return processing. The chapter ends with a discussion of mixing in the box and some suggestions for maximizing your results when mixing in Logic Pro X.

⊕ Learning Targets for This Chapter

- Set effective levels for your tracks
- Recognize the role of EQ and dynamics processing in creating a balanced mix
- Use panning to create a sense of space and positioning in your mix
- Understand how inserts and sends can be used to add processing to tracks
- Recognize the advantages of in-the-box mixing

Key topics from this chapter are illustrated in the Logic Audio Production Basics Study Guide module available through the Elements|ED online learning platform. Sign up at ElementsED.com.

After you have recorded, imported, or otherwise created the media that you want to use in your project, you can set about the task of mixing the audio. Mixing is the process of setting the basic levels, positioning, and sonic characteristics for your tracks. Essentially, this is where you determine how the various parts of the project will blend together to create a cohesive result during playback.

Basic Mixing

Although mixes can get very complex on large projects, the basic process is fairly simple. The primary goal when creating a stereo mix is to set the levels for each of the tracks using the corresponding volume faders and to position each of the tracks' signals within the stereo field using pan controls. Other options in Logic Pro X include using inserts and sends to process signals and add effects.

Setting Levels

Setting the levels for your tracks is typically done using channel strips in the Mixer. You can display the Mixer in the main window or open it as a separate window (**WINDOW > OPEN MIXER**). Here you'll find channel strips for each track in your project. You can use the volume faders for each channel strip to adjust each track's overall output level.

Figure 7.1 Channel strips in Logic Pro's Mixer window

Level Considerations

The purpose of adjusting the output level of a track is to make the track audible without obscuring other tracks or causing the track to become overly prominent in the mix. During the recording process, audio is often recorded louder than it needs to be in the final mix. Additionally, as you begin summing multiple tracks together, the overall output level for the project will increase. For these reasons, it helps to pull all of the faders down quite a bit when you begin mixing and then gradually adjust the levels of each track as needed.

Allowing one track to be heard above all the others is not only a matter of raising its volume fader, however. A common mixing dilemma is that making one track louder will cause another, equally important track to become overshadowed and indistinct.

> **Getting a Good Music Mix Is More Than Setting Levels**
>
> Raising the fader on a guitar channel strip enough to get it to cut through the mix can make the vocal hard to hear. Raising the fader on the vocal channel strip may then begin to obscure the impact of the drums. And raising the faders on the drum channel strips can affect the bass guitar, which may now be barely audible. By the time you bring up the level on the bass guitar to compensate, you may be right back where you started, with the guitar no longer cutting through the mix. Only now, everything is louder!

In the quest for the ultimate mix, where "everything is louder than everything else," often you simply end up with a very loud mix and still cannot distinguish individual parts.

As you start setting the initial levels for your mix, try to achieve a good overall balance. Don't get overly concerned if you find that certain tracks start to compete with one another. These issues can be addressed later using panning, EQ and dynamics processing, and other techniques that help give each track its own unique space in the mix.

Beware of Clipping

Among the problems caused by loud tracks, the output levels of your mix may get too "hot," leading the signal to clip at the digital–to–analog converters. This will cause the mix to become distorted on playback during the loudest moments.

Digital distortion is always detrimental to sound quality and should be avoided at all costs. Therefore, it is better to mix too quiet than too loud. You can always increase the overall output levels of a mixed stereo file at a later stage if needed. But you cannot remove clipping from the stereo file after the fact.

Metering on Channel Strips

To help you keep an eye on your audio levels, each channel strip includes a standard meter display to the right of the track fader. At the top of each meter is a peak level display. This display will light red whenever the signal from the track exceeds full-scale audio.

 Meters in the Logic Pro mixer measure loudness in decibels relative to full-scale audio (dBFS), with full scale represented by 0 at the top of the meter. Digital clipping occurs whenever a signal exceeds 0 dBFS at an input or output.

Figure 7.2 Stereo output meter with peak indicator shown

Audio channel strips show *post-fader* metering by default in Logic Pro X. This means that the meter reflects the output level of the track after the channel strip rather than the level from the source audio on the track (disk file). A lit peak indicator on an Audio channel strip means the output levels are too hot—the volume fader should be pulled down, reducing the playback amplitude.

 Track metering reflects plug-in processing done on the track's channel strip, including changes made by EQ or dynamics processors. Any boost or cut to the signal caused by plug-ins will be represented on the meter.

Metering on Output Channel Strips

To control the output levels of your mix, you can use the Master Volume slider in the main window control bar. You can also use the volume fader on the master channel strip in the mixer. However, the master channel strip does not include a meter. To monitor your output levels, you must reference the meter on the output channel strip. By default, a project contains one output channel strip, called **Stereo Out**. This channel takes the signal from all the individual tracks in the project.

The level meter on the **Stereo Out** output channel strip reflects the overall output of the entire mix. This is a summed total of the signals from each contributing track.

Peak indicators on the output channel strip light red when the summed signal level exceeds the capabilities of the digital-to-analog converters. In this case, you can reduce the levels on each of the contributing tracks in your mix, or reduce the overall output level using the volume fader on the output channel strip itself.

While working on your mix, keep an eye on the meters on the **Stereo Out** channel strip. If it begins clipping, reduce the output levels before continuing.

The Role of EQ and Dynamics Processing

As mentioned above, setting the fader levels is not the only consideration when it comes to allowing each track to be heard in a mix. Other important considerations include the track's frequency spectrum and the track's dynamic response.

The frequency spectrum of a track, or the amount of energy the track has at different audio frequencies, can be shaped using equalization (or EQ for short). You can use EQ processing to help reduce low-end rumble, high-end hiss, and other unwanted noise in a signal. You can also use EQ to shape a signal and create a unique tone for each track. This allows certain target frequencies or frequency ranges to be more prominent than others. By selectively boosting and cutting frequencies in each track's channel strip, you can balance the audio spectrum in the mix, and give each track its own sonic footprint.

In the interest of keeping track levels in check, it is generally better to focus on cutting the frequencies you don't want rather than boosting the frequencies you wish to emphasize.

The dynamic response of a track refers to the range of amplitude values on the track. Said another way, the track's dynamics relate to the differences in loudness from one part of the track to another. The momentary loudness peaks (or *transients*) on a track are often much louder than the track's average levels. These loudness peaks can begin to obscure other tracks (or cause clipping) long before the source track's overall amplitude is where you want it in your mix.

A solution to this problem is to use compression on the track's channel strip. Compression helps reduce the loudness in peak areas without affecting the average loudness. Using compression in combination with an appropriate amount of make-up gain can help you increase the level from a track without obliterating the other tracks in your mix.

For details on using EQ and dynamics processing, see the associated discussions in Chapter 8 of this book.

How to Set Levels

When it comes to setting the relative levels for your tracks, it helps to get familiar with the audio characteristics of each track first. Then consider how those characteristics contribute to the overall mix. Use the following steps as a guide to get started:

- **Listen to the track in isolation**—Use the Solo function (**S** button in the channel strip) to isolate each track in turn and familiarize yourself with its sonic characteristics. When working on a music mix, consider how the track supports the rhythm, groove, harmony, and melody of the piece. When working with non-musical material, consider how the track contributes to the tone and emotion of the mix, and listen for the clarity of the track.

- **Listen to the track in context**—Unsolo the track to determine the appropriate level for the track relative to the other tracks in the mix. Can you still hear the important sonic elements that you heard when the track was soloed? Or is the track completely lost among competing sounds?

- **Listen to the mix without the track**—Mute the track (**M** button in the channel strip) to gauge its contribution to the mix and how necessary it is. Does the mix sound lacking without the track? Or is the mix suddenly clearer without the track competing for sonic space?

- **Check the track throughout the piece**—Toggle the mute and solo state of the track on/off to hear the track isolated, in context, and removed from the mix in multiple locations throughout the composition. Are there some parts of the mix where the track is more important than others? Should the levels change for different sections of the mix?

- **Make adjustments in iterations**—Return to the track and adjust the levels as you make other changes to the mix. Setting levels is an iterative process that will require fine-tuning as your mix begins to take shape. Be sure to consider how each track's contribution changes as you begin setting levels for other tracks, adjusting pan positions, and adding processing to the mix.

Work your way though each track in the project, listening and adjusting levels as you go to create a good overall balance. Don't worry if there are aspects of the mix you aren't completely happy with at this point; you will be refining the result as you progress through later stages of the mixing process.

Panning

The pan controls on each track's channel strip allow you to position the track within the stereo field. In a stereo mix, you will have two track formats to consider: mono tracks and stereo tracks.

Pan Controls on Mono Tracks

Mono tracks will have a single pan/balance control that allows you to position the track's output as desired in the stereo spectrum. Panning a track hard left will cause it to play out of the left speaker only, panning to center will cause the track to play at equal volume out of both speakers, and panning hard right will cause the track to play out of the right speaker only.

Figure 7.3 Mono tracks panned hard left, center, and hard right (from left to right)

Pan Controls on Stereo Tracks

Stereo tracks will have a few panning options: Balance, Stereo Pan, and Binaural Pan. Here we focus mainly on Balance Pan knobs and Stereo pan knobs. You can switch between these panning options by Control-clicking on the pan knob and making a selection from the pop-up menu.

Binaural—Binaural panning simulates spatial information like the elevation and distance of a sound. This kind of processing is best suited for headphone playback.

Balance—Balance panning is the default panning mode for stereo channel strips. It processes the relative left and right channels together allowing you to position the entire track within the stereo field.

Stereo Pan—The Stereo Pan knob allows you to position the left and right signals independently within the stereo field.

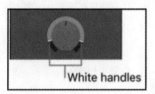

Figure 7.4 Stereo Pan Knob with left and right white handles for adjustment

When using the stereo pan knob, you can adjust the stereo spread of one channel independently of the other. To narrow the stereo field for the right channel only, drag the white handle of the right side closer to center position. You can invert the left and right channels by Command-clicking inside the pan knob or ring. You'll see the ring color switch from green to orange. Command-clicking a second time will reset the channels.

You can position the stereo spread by dragging the center of the pan knob vertically. To reset these values to their starting positions you can Option-click anywhere in the pan knob or ring.

Figure 7.5 Stereo tracks with default balance pan knob (left), stereo pan knob (center), and inverted stereo panning (right)

Panning Examples

As you refine your mix, you'll want to consider the panoramic position of each track, along with the overall distribution of the tracks. Positioning individual tracks at distinct locations will give each track its own space in the mix and help give your mix width and realism.

- **Center–panned tracks**—Certain tracks tend to work best in the center of a mix. These typically include the tracks that provide the main focus or lead part for the piece, such as a narration or voiceover track or a lead vocal track. In music production it is also common to place the main rhythmic elements at or near the center of the mix, such as the kick drum, snare drum, and bass guitar.

- **Hard–panned tracks**—Tracks that provide flavor, color, or character to a mix are often effective when panned hard left or right. These are typically supplemental tracks such as a shaker or harmonica in a music mix or ambient accents such as a distant siren or muffled argument from an adjacent room in a fiction narrative.

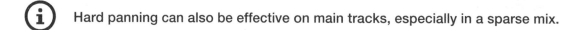 Hard panning can also be effective on main tracks, especially in a sparse mix.

One potential problem with using hard-panned tracks is that they can become inaudible to a listener who is on the opposite side of the listening space from the sound source, such as when the stereo speakers are placed wide apart in a living room. Also, if the stereo playback should drop a channel, everything hard-panned to the dropped side would be lost.

These problems can be mitigated by not panning fully left/right, by including some reverb from a hard-panned track in the opposite channel, and by using psycho-acoustic processing to place an image in a stereo field by means of timing offsets while keeping the signal in both channels.

 For an example of hard panning, check out the electric guitar on early albums by Van Halen. Hard panning is prominent in songs such as "Little Dreamer," "Beautiful Girls," "Hang 'Em High," and numerous others.

- **Tracks panned off–center**—Often tracks are placed off center to create realism and a sense of space in the mix. This technique is commonly used to position characters around a room in a dialog mix. In music mixing, it is common to pan individual drum tracks to emulate the layout of the drum kit (placing hi-hat and splash cymbal on the left, rack toms across the middle, and floor tom and ride cymbal on the right, for example). Another common technique is to pan backing vocal tracks, placing high harmonies on the left, mid-range parts near the center, and low harmonies on the right, for example.

- **Other ideas**—It can be very effective to offset parts that have similar impact in a mix. For example, panning a doubled vocal part to left and right extremes can create a sense of width in the mix.

Similarly, panning a rhythm guitar part opposite a Rhodes piano part can help each part contribute equally without competing with one another.

 For examples of effective panning for multiple vocal parts, take a listen to "Ziggy Stardust," "Changes," and other classics by David Bowie. Bowie often used opposite-panned vocal doubles and harmonies during key sections for emphasis.

Staying Organized

By default, Logic Pro colors tracks and regions in your project by track type. It can be helpful to give them more meaningful colors to visually organize your project.

Using the Color Palette

You can assign color to channel strips or regions by accessing the color palette. To display the color palette, choose **View > Show Colors** or press **Option+C**.

Figure 7.6 The color palette in Logic Pro X

With the Color Palette open, you can apply colors to tracks and regions depending on what's selected. If one or more tracks in the main window are selected, you can choose a color to apply to regions on the tracks. If you select a channel strip from the mixer and choose a color, color will be added to the channel strip/track but not to regions.

Track color indication is shown to the left of tracks in the main window and on the bottom of channel strips in the mixer.

Using Menu Commands

When the color palette is not open, you can assign colors using a command under the Track menu or various Control-click menu commands.

To change track colors, do one of the following:

- Select **Track > Assign Track Color**, then select a color from the color palette.

- **Control-click** on the track header, choose **Assign Track Color** from the pop-up menu, then select a color from the color palette.

- **Control-click** on the channel strip in the Mixer, choose **Assign Channel Strip Color** from the pop-up menu, then select a color from the color palette.

Figure 7.7 Assigning colors from the Control-click menus: main window (left) and mixer (right)

Region colors can be changed to follow the track color, or they can be customized using the color palette.

To change a region color, do one of the following:

- **CONTROL-CLICK** the region and select **NAME AND COLOR > COLOR REGIONS BY TRACKS** to automatically color the selected regions to match track/channel strip colors.

- **CONTROL-CLICK** the region and select **NAME AND COLOR > SHOW COLOR**, then make a selection from the color palette.

Processing Options and Techniques

When it comes to adding processing to your tracks, you have two broad categories to consider: gain-based processing and time-based processing. You also need to consider when and how any processing should be added to your mix. Here we'll cover some options available in Logic Pro X.

Gain-Based Processing

Gain-based processors include any processors that affect the amplitude of the audio signal in some way. Examples include EQs, compressors, noise gates, expanders, and similar dynamics processors.

When adding gain-based processing to a signal, you will typically assign the processor as an insert to an individual track. This type of processing is usually applied to the entire signal (100% wet), rather than being mixed together with a dry signal. The processing is also commonly track-specific.

For example, when using an EQ plug-in to eliminate low-frequency rumble, you will want 100% of the source signal to be affected. And you will adjust the EQ parameters to address the specific frequency characteristics of the source signal on that individual track.

Time-Based Processing and Effects

The other type of processing can broadly be classified as time-based processing and effects. This includes processors that affect the signal in the time-domain, such as reverbs and delays, and modulation effects, such as choruses, flangers, and phasers. These processors typically get applied to only a portion of the signal and are mixed back in with the dry signal.

Time-based processors are commonly used as shared effects, across multiple tracks in a mix. This helps provide consistency in the mix and makes more efficient use of your processing resources.

Insert Effects Versus Send Effects

Logic Pro X allows you to add signal processing to a track's channel strip using Insert slots (also known as Audio Effect slots) and Send slots. An Insert slot is an audio patch point that places a signal processor directly into the signal path of the channel strip. When using an insert effect, all of the audio from the track must pass through the effect on its way to the track's output.

Logic Pro X provides 15 Insert slots on each audio channel strip, allowing you to process the signal through multiple successive plug-ins, as needed.

See Chapter 8 in this book for details on using Audio Effect slots for plug-in processing.

By contrast, a Send slot provides a signal path that can be used to route audio from one or more source channel strips to a parallel destination for effects processing. In Logic Pro X, whenever you activate a Send slot by selecting an unused bus, an aux channel strip is created for that bus. Effects processing is commonly added to the signal using an Audio Effect plug-in inserted on the aux channel strip.

In this send-and-return configuration, the aux channel strip is commonly referred to as the *return* channel.

When creating a send effect, you use a bus to route signals from the source track(s) to the destination aux channel strip. You use the Send Level knob for the Send slot to control the signal level being sent to the bus.

Processing with External Gear

In Logic Pro X, you can also choose to route a signal from a Send slot through external hardware, provided you have sufficient I/O on your audio interface. In this case, the send will route out of your audio interface, through an external processor, and back into your audio interface before the signal is returned to the mix. (See Figure 7.5.) As above, the signal is returned to the mix using an aux channel strip.

Figure 7.8 External reverb connected to an audio interface for use with Logic Pro X

Mixing in the Box

Although it is possible to use external gear for send-and-return processing in Logic Pro X, mixing completely "in the box" offers numerous advantages. Mixing in the box simply means that all signal processing is provided by internal software plug-ins, so the mix does not rely on any external gear.

Advantages of In-the-Box Mixing

The advantages of working completely in the box include the following considerations:

- **Portability**—Creating your Logic Pro X mix entirely in the box means that you can take the project file out of your studio and open it from any Logic Pro workstation anywhere in the world. Barring any missing plug-ins, the mix will sound the same from any system.

  If you were to use external gear, you will need to take the gear with you to work on the mix from a different location.

- **Recallability**—Saving the project will save all plug-in settings and levels, allowing them to recall exactly as they were the next time you open the project. This is not the case when using external gear, as you will need to reconfigure all settings on the external hardware to match your last used settings for the project.

- **Time savings**—Reconfiguring hardware is not only inconvenient, it can be quite time-consuming. It also requires keeping detailed notes of settings and configurations, and updating those notes every time you change a setting.

- **Dynamic automation**—Plug-in settings can be automated, enabling you to incorporate dynamic changes in your mix. All automation settings are saved and recalled with the project, ensuring that it plays back consistently every time.

Getting the Most Out of an In-the-Box Mix

To get the most out of an in-the-box mix, you will want to consider how much control you need to have over the sonic characteristics of your mix. Ideally, this is something you would begin thinking about at the project inception, prior to starting to record.

For example, you'll need to decide early on if you want to be able to control the dynamics of each track entirely from within the project, or if it is OK to apply some compression to individual signals prior to the recording input stage. Likewise, you'll need to decide whether to record sources such as guitar tracks from an amp, with the guitarist's effects processing chain already in place, or to record a direct signal from the guitar and add amp simulators and stomp-box effects as plug-ins inside of Logic Pro X.

Aside from considerations about when to apply processing, you will also need to take steps to maximize the results of any processing you apply from within Logic Pro X. Use the following simple suggestions as a starting point:

- **Learn how to set plug-in parameters**—To effectively use the plug-ins that are provided with Logic Pro X, you will need to know how to set the parameters properly and recognize what each parameter is used for. Take time to experiment and practice using each plug-in on material you are familiar with.

- **Avoid using copies of the same plug-in on many different tracks**—Learn how to use send-and-return processing to apply the same effects to multiple tracks, rather than placing individual copies of the same plug-in on each track's channel strip. Not only does using multiple copies of a plug-in require more processing power than using a single copy, but keeping the settings in sync across multiple tracks can become tedious and time-consuming.

- **Invest in good quality plug-ins**—The plug-in collection that comes with Logic Pro X is an excellent starting point. As you expand beyond that set, invest in quality processors that enhance your sonic palette. Don't be afraid to spend some money to get what you want, but keep in mind that many high-quality plug-ins are available at very reasonable prices. Spend some time reading or viewing product reviews, and check out trial versions, if available. Once you know what you want, keep an eye out for discounts and sale prices.

> Plug-in manufacturers commonly offer discounts and promotional pricing during national holidays, trade shows, and similar events.

Review/Discussion Questions

1. What are some reasons for lowering the faders across your entire project as you begin to set levels for your mix? (See "Level Considerations" beginning on page 168.)

2. Why is it important to avoid clipping in your mix? How can you tell if the project is clipping at the outputs? (See "Beware of Clipping" beginning on page 169.)

3. How can equalization be used to help a track cut through a mix? (See "The Role of EQ and Dynamics Processing" beginning on page 171.)

4. What is meant by the dynamic response of a track? What can be done to "tame" excessively loud peaks on a track? (See "The Role of EQ and Dynamics Processing" beginning on page 171.)

5. How are the pan controls different between mono tracks and stereo tracks? What options are available for the pan controls on stereo tracks in Logic Pro X? (See "Panning" beginning on page 172.)

6. Describe some examples of panning techniques that can be useful or effective in a mix. (See "Panning Examples" beginning on page 174.)

7. How does gain-based processing affect an audio signal? Give some examples of gain-based processors. (See "Gain-Based Processing" beginning on page 176.)

8. What are some available time-based processors? How are time-based processors typically applied to tracks? How is this different from the way gain-based processors are used? (See "Time-Based Processing and Effects" beginning on page 177.)

9. How are insert effects different from send effects, in the way they process audio? What is the role of an aux channel strip in a send-and-return configuration? (See "Insert Effects Versus Send Effects" beginning on page 177.)

10. What is meant by "in-the-box" mixing? What are some advantages of mixing in the box? (See "Mixing in the Box" beginning on page 178.)

11. Why is it important to begin thinking about in-the-box mixing from the project inception? (See "Getting the Most Out of an In-the-Box Mix" beginning on page 179.)

12. How can you maximize the results of any processing you apply from within Logic Pro X? (See "Getting the Most Out of an In-the-Box Mix" beginning on page 179.)

To review additional material from this chapter and prepare for certification, see the Logic Audio Production Basics Study Guide module available through the Elements|ED online learning platform at ElementsED.com.

Exercise 7

Creating a Basic Mix

🎧 Activity

In this exercise, you will perform basic mixing tasks in Logic Pro X. You'll start off by adjusting levels to get a good basic mix. Then you'll explore some strategies to prevent clipping on loud tracks. Finally, you'll adjust pan settings to enhance the mix and create space for different instruments.

🕒 Duration

This exercise should take approximately 20 minutes to complete.

◈ Goals/Targets

- Set levels for different tracks in the project
- Set panning for different tracks in the project
- Save your work for use in Exercise 8

Exercise Media

This exercise uses media files from the song, "Lights," provided courtesy of Bay Area band Fotograf.

Written by: Zack Vieira and Eric Kuehnl; Performed by: Fotograf

The media provided for this course may be used for educational purposes only. No rights are granted to use the media for any other personal, commercial, or non-commercial purposes.

Getting Started

To get started, you will open your completed project from Exercise 6. This will serve as the starting point for this exercise.

Open your existing Lights project:

1. Select **FILE > OPEN** or press **COMMAND+O** (Mac). The Open dialog box will display.
2. Navigate to the location where you saved your project in Exercise 6.
3. Select your **Lights-xxx** project and click **OPEN**. The project will open as it was when last saved.
4. If needed, choose **VIEW > SHOW MIXER** or press **X** to open the Mixer tab.
5. Choose **FILE > SAVE AS** and create a copy of the project called **Lights02-xxx** keeping the project in the original project folder.

Creating a Rough Mix

At this point, you're ready to start mixing. As you adjust the volume faders on the channel strips in your project, you will want to keep an eye on your output levels.

Preparing to Mix

Before you get started mixing it is a good idea to get your tracks organized. You'll want to make sure the tracks are in an order that makes sense. You may also want to color tracks and regions for visual organization.

Arrange the track order in the main window by clicking and dragging on track nameplates:

1. Move the **07 Loops** track after the **04 Claps** track so that all the drums and percussion are together.
2. Move the **08 Bass Gtr** track after **05 Synth Bass** track so that the bass tracks are together.
3. Move the **06 Vox** track all the way to the bottom so that it is separated from all the other tracks.

Color-code your channel strips:

1. Right-click on any channel strip. From the pop-up menu, select **TRACK HEADER COMPONENTS > TRACK COLOR BARS** to display color bars in the Tracks area.
2. Open up the color palette by selecting **VIEW > COLOR** or pressing **OPTION+C**.
3. Click on the **01 Kick** channel strip to select it, then **SHIFT-CLICK** on the **07 Loops** channel strip to extend the selection across all of the drums and percussion.

4. If regions are selected in addition to the channel strips, click any empty space in the Tracks area to deselect the regions while maintaining channel strip selections.

5. Select a yellow swatch (or another color of your choice) from the palette to color all the drum and percussion tracks the same color.

6. Select the **05 Synth Bass** and **08 Bass Gtr** tracks.

7. Again click in an empty space, as needed, to deselect all regions on the tracks; then click an orange swatch (or color of your choice) from the palette to make the bass tracks the same color.

8. Click on **05 Synth Pads** then shift click on **Bright MK II Blackface** to select all the keyboard instruments.

9. Deselect any regions on the track; then choose pink (or color of your choice) from the palette for the three selected tracks.

10. Click on the **Smart Strings** track, deselect it regions, and select purple (or color of your choice) from the color palette.

11. Click on the **06 Vox** track, deselect its regions, and select a light blue (or color of your choice) from the color palette.

Once that's done, you can color-code regions to match the track colors.

Color-code regions:

1. In the main window, select all the regions by pressing **COMMAND+A**.

2. **CONTROL-CLICK** on one of the selected regions and select **NAME AND COLOR > REGIONS TO TRACK COLOR**, or press **OPTION+SHIFT+C**.

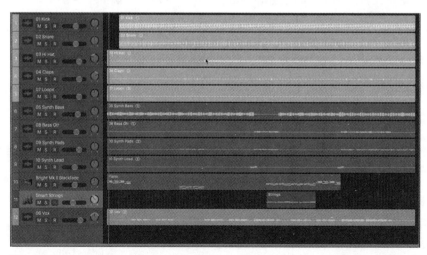

Figure 7.9 The tracks in the main window, showing organization by instrument and color-coding

General Mixing Suggestions

Here are some general tips for setting the rough levels for tracks:

- Solo a track and listen closely to the content. What are the most important details of that instrument? When you unsolo the track can you still hear those details clearly in the overall mix? If not, you may need to increase the level of the track.

- Mute a track and listen to the mix. Does the mix sound empty without that track? Do you even notice that it is missing? You may decide to leave the track muted or reduce the level of the track.

- While listening through the whole song, consider whether a track is only necessary in a particular section of the song. Most songs build in complexity as they progress. You might want to consider muting or reducing the volume of a track at the beginning of the song or during verses. Then you can unmute or increase the volume of the track later in the song or during choruses.

- Keep revisiting the various sections of the song. As you make changes in one section, it may change how you feel about other sections.

Suggestions for This Project

Here are some specific suggestions to help you create a rough mix of your project.

- Listen to the mix going into Chorus 1 (Bar 48). The **10 Synth Lead** and **Bright MK II Blackface** tracks are a bit too loud in these sections. Try reducing the level of each track by about –5 dB.

- Listen to the mix in Chorus 2 (Bar 77). The **Smart Strings** track is much too loud. Try reducing the level on this track by around –8 dB.

- Throughout the song, the vocal track (**06 Vox**) gets a little bit lost in the mix. Try increasing the level on this track by around +1.5 dB.

- The **08 Bass Gtr** track gets a little too loud during both choruses. Try reducing the volume on this track by about –3 dB.

- The **01 Kick** and **02 Snare** tracks are both a little loud throughout the song. Try dropping them both by about –4 dB.

Adjusting Levels to Prevent Clipping

When mixing a project, it is fairly common to have some tracks that show clipping (or "overs") on individual track meters or on the Stereo Output. This can be tricky to deal with when you're first developing your mixing skills. Fortunately, you have a number of ways to address clipping in Logic Pro X.

General Level Suggestions

Here are some tips for dealing with clipping:

- If an individual track is causing clipping at the outputs (as displayed on the Stereo Output channel strip), the first thing to do is reduce the level from the associated track. If you can prevent clipping while the resulting level still sounds good in the mix, you're all set. If the resulting level is too quiet relative to other tracks, you'll need to try another approach.

- Another option is to reduce the levels of all of the tracks. To do this, first reduce the level of the track that is causing the clipping until the clip goes away. Take note of how much you reduced the track from its prior level in the mix (such as –6 dB). Then reduce the level of each track in the project by the same amount.

- A third option to prevent clipping is to simply lower the fader on the master channel strip (or Stereo Output channel strip) to correct the clip at the output stage. While clipping on an individual track may not be detrimental to the final mix, you never want to have clipping on the Stereo Output channel strip.

Suggestions for This Project

The individual tracks in this mix are well below clipping, so you should not need to reduce them further. But the overall mix is clipping. You'll use the Stereo Out channel strip to solve this problem.

Correct clipping at the output stage:

1. Play through the song.

2. Keep an eye on the **Stereo Out** channel strip, watching the level on the meters and the peak level indicator above the meters. (See Figure 7.10.)

3. Take note of the loudest areas of the project. After playing the song all the way through, you should have an idea of where the signal is peaking.

4. Reduce the level of the **Stereo Out** track by at least the amount shown in red in the peak level display. (Alternatively, if the peak is below zero, you could raise the level until the peak is just below clipping.)

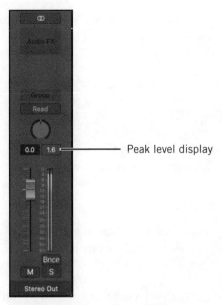

Figure 7.10 The peak level display (red) on the Stereo Out channel strip, just under the pan knob

5. Click on the peak level display to reset it.

6. Play through the song again to verify that the peak levels now remain below clipping. If clipping still occurs, reduce the Stereo Out level further and test again.

Adjusting Panning

In this part of the exercise, you'll adjust the panning for your tracks. The project includes both mono and stereo tracks. Mono tracks can be easily positioned to any location in the stereo field. Stereo tracks are often adjusted to have a narrower or broader width. Both of these techniques can be used to add interest to the mix, and can also create space where each track can be heard and appreciated.

General Panning Suggestions

Here are some general tips for panning tracks:

- For each mono track, solo the track and adjust the panning. Listen for a particular pan position that just "feels right" for the track or instrument. Unsolo the track to check how its position works in the overall mix.

- For each stereo track, try using the stereo pan knob to adjust the panning so that the width is narrower. Control-Click on a pan knob to select stereo pan. You can often create more space in the mix by narrowing some stereo tracks. See if any tracks sound better left at the default (widest) pan setting instead.

- Try panning some tracks in unusual ways. Mono tracks can be panned all the way left or right to create a novel sound. Stereo tracks can be shifted so that they are not symmetrical in the left and right channels. Just be careful because extreme panning can also distract from the listening experience.

Suggestions for This Project

Here are a few specific options you can try for altering the panning of your project:

- Try setting the **03 Hi Hat** channel strip slightly to the left (–25) and the **04 Claps** channel strip slightly to the right (+25). This can help to create a more interesting mix of the percussion elements. (You can hear the effect of these changes in each of the choruses, where both tracks are contributing to the mix.)

- Create some space for the vocal by narrowing the panning on the **06 Vox** channel strip a bit. **Control-click** and select **Stereo Pan**, then try setting the left and right white handles to around the 10 o'clock and 2 o'clock positions, respectively.

- Pan the **Smart Strings** channel strip slightly to the left to give the vocals more space (–15). (You can hear the contribution of this track in Chorus 2 at Bar 77.)

Finishing Up

To complete this exercise, you will need to save your work and close the project. You will be reusing this project in Exercise 8, so it's important to save the work you've done.

Before you wrap up, you can also listen to the project to hear the mixing changes you've made.

Finish your work:

1. Choose **File > Save** to save the project.

2. Press **Return** to position the playhead at the beginning of the project.

3. (Optional) Press the **Spacebar** to begin playback and listen through the project. Press the **Spacebar** a second time when finished to stop playback.

4. Choose **File > Close Project** to close out of the project.

That completes this exercise.

Chapter 8

Signal Processing

...*What You Can Do to Optimize Your Audio*...

This chapter begins with an overview of using plug-ins in Logic Pro X. We take a look at some general categories of plug-ins—including EQ, dynamics, and time-based effects—and provide some suggestions for using various plug-in processors and parameters. We also look at the process of creating send-and-return configurations and discuss how this setup can be useful.

⊕ Learning Targets for This Chapter

- Understand basic plug-in functionality in Logic Pro X
- Become familiar with the major plug-in categories
- Learn how to use basic plug-in parameters
- Explore send-and-return configurations

Key topics from this chapter are illustrated in the Logic Audio Production Basics Study Guide module available through the Elements|ED online learning platform. Sign up at ElementsED.com.

Plug-ins are the key to one of the most important aspects of working with a DAW: processing audio and adding effects. Innovations in plug-in design were a key factor in the rise of the DAW as the dominant platform for producing music and creating audio for film, TV, and video games. Without great-sounding plug-ins for equalization, dynamics processing, and time-based effects such as reverb and delay, the DAW revolution might never have occurred.

In this chapter, we take a look at how to use plug-ins in Logic Pro X and review the most common types of plug-ins used in today's production workflow.

Plug-In Basics

The sheer variety of plug-ins available today can seem completely overwhelming when you're just getting started. Logic Pro X comes with a large collection of professional effects that are capable of handling the most common production needs, as well as a number of specialty plug-ins. While the selection is vast, with practice, you'll eventually become familiar with these tools and how to use them effectively.

For those who are new to the DAW world, learning just a few plug-ins well is a realistic goal. And, because there are only a few major categories of plug-ins, the knowledge that you gain now can easily be applied to other plug-ins down the road as you expand your processing palette.

Inserting Plug-Ins on Tracks

We've already looked at several ways to insert plug-ins (including virtual instruments) on tracks in Logic Pro X. But let's do a quick recap before moving on to using plug-ins to process audio.

To insert a plug-in on a track in the Inspector:

1. Click the **INSPECTOR** button (or press I) to open the Inspector.

2. Select the desired track from the main window.

3. Click on the Audio Effect slot for the desired insert position.

4. Select the desired plug-in from the pop-up menu. The plug-in will be applied at the selected position, and the plug-in window will open.

To insert a plug-in on a track in the Mixer:

1. Press the **MIXER** button (or press **X**) to open the Mixer. Alternatively, you can open a separate Mixer window by choosing **WINDOW > OPEN MIXER** (or press **COMMAND+2**).

2. Adjust the size of the Mixer or scroll vertically to bring the audio effect section into view.

3. Click on the Audio Effect slot for the desired insert position on the track's channel strip.

4. Select the desired plug-in from the pop-up menu. The plug-in will be applied at the selected position, and the plug-in window will open.

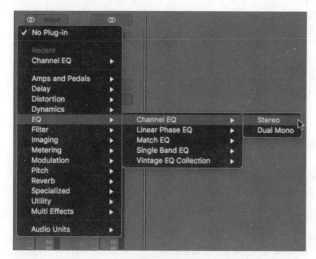

Figure 8.1 Assigning the Channel EQ plug-in to an Audio Effect slot

> ⓘ Some audio effects will have multiple output options, like stereo or dual mono. Others may have only one option. If you select a plug-in without picking the output preference, Logic Pro X will automatically select the first option.

> ⓘ In a new project, a single Audio Effect slot will be visible on each track. The audio effect section will expand to show additional slots as needed. The last visible empty slot will be half the normal size. (See Figure 8.2.)

To change or replace a plug-in on a track:

1. Move the pointer over the plug-in slot and click the arrows that appear on the right side.

Figure 8.2 Moving the pointer over the plug-in slot reveals the arrows to open the pop-up menu

2. Make a new plug-in selection.

To remove a plug-in from a track:

1. Move the pointer over the plug-in slot and click the arrows that appear on the right side.
2. Choose **NO PLUG-IN** from the pop-up menu.

Moving and Duplicating Plug-Ins

After you've begun working with plug-ins on tracks in a project, you may find that you need to move a plug-in to a different insert slot or to a different track entirely. You may also want to create a duplicate of the plug-in elsewhere in the project. These tasks are easily accomplished in Logic Pro X.

To move a plug-in, do the following:

- Click on the plug-in and drag it to a new Audio Effects slot on the same track or to an available Audio Effects slot on another track.

To duplicate a plug-in, do the following:

- OPTION-CLICK on the plug-in and drag it to the desired location (on the same channel strip or on another channel strip). A duplicate plug-in will be added at the destination location.

> You can move or duplicate a plug-in within the same channel strip using the Inspector or Mixer. If you want to move or duplicate a plug-in to a different channel strip, you must use the Mixer.

Displaying Plug-In Windows

Before you can adjust the parameters of a plug-in, you'll need to open the plug-in window. Plug-in windows open automatically when you assign a plug-in to a slot. You can easily re-open a plug-in window that is not displayed at any time.

To open a plug-in window:

- Click on the center of the plug-in slot on the track.

Figure 8.3 Opening the plug-in window by clicking on the center of a plug-in slot

Common Plug-In Controls

Logic Pro X plug-ins feature two sets of controls: the standard Logic Pro X controls that are common to all plug-ins, and the process-specific controls that are unique to each plug-in. Let's take a look at the common controls first. We'll look at process-specific controls later in this chapter.

The common controls are available in the plug-in window header.

Figure 8.4 The plug-in window header shared by all AU plug-ins in Logic Pro X

Bypass Control

To the right of the Automation controls is the plug-in Bypass button.

Figure 8.5 Plug-in Bypass control

Use this button to disable the currently displayed plug-in so that it will not affect the track's audio signal. This allows you to perform an "A/B" comparison of the affected and unaffected signal.

Plug-in Settings Controls

To the right of the bypass button are the plug-in Settings. These controls are used to select and manage presets for the plug-in. This area features a number of items: the Plug-In Settings menu, Previous and Next Setting buttons, the Compare button, the Copy and Paste buttons, and the Undo and Redo buttons.

Figure 8.6 The plug-in Settings controls

Plug-in Settings controls include the following:

- **Plug-In Settings menu**—Use this pop-up menu to select previous/next settings, copy/paste settings, save/load settings, delete settings, and select a setting from the submenus.

- **Previous and Next settings buttons**—Click the left or right arrow buttons to quickly select the previous or next preset.

- **Compare button**—Click this button to toggle between the most recently loaded preset and any subsequent modifications that you've made. This is often referred to as an "A/B" comparison.

- **Copy and Paste settings buttons**—Use these buttons to copy and paste settings between different plug-in instances.

- **Undo and Redo settings button**—Use this button to undo and redo setting adjustments.

View Control

To the right of the Settings controls is the View pop-up menu.

Figure 8.7 Plug-in View pop-up menu

Use this pop-up menu to adjust the size of the plug-in window. Alternatively, you can adjust the size of the window by dragging any corner. The menu also allows you to switch between plug-in parameter views:

- **Controls**—This view shows all parameters as numerical fields with horizontal sliders. You can adjust values by moving the sliders or entering values into the numerical fields.

- **Editor**—This view shows the plug-in graphical interface. (This is usually the default view if a graphical interface is available.)

 The plug-in automation features in Logic Pro X are not covered in this book. See the Logic Pro X User Guide for information about plug-in automation.

Link Control

When Link is enabled, a single plug-in window is used onscreen as you open new plug-ins. The plug-in window updates to display the newly opened plug-in. By default, Link is disabled.

Figure 8.8 Plug-in Link control

Right-click or Control-click the Link button to access the menu and choose from the following:

- **Off**—Plug-in windows are not linked. Clicking on the center of the plug-in slot will open the plug-in in a new window. As additional plug-ins are opened, multiple windows will be visible.

- **Single**—A single plug-in window is used to display any opened plug-in on any track. The window will refresh to reflect the current selection. When enabled, the link button will turn purple.

- **Multi**—A single window is used for plug-ins on the same slot row of any track; plug-ins on a different row will open in a separate window. Additionally, the plug-in window for a slot row will refresh to show the associated plug-in for any newly selected track. When enabled, the link button will turn yellow.

> ⓘ Link provides a powerful way to see and adjust plug-in settings across your project. If you display multiple plug-ins from track 1 and then select track 2, the open windows will update to display the associated plug-ins from track 2.

Close Control

The icon on the top-left of the plug-in window can be used to close the plug-in.

Figure 8.9 Close plug-in window icon

Show/Hide Control

On the right of the plug-in window is the Show/Hide button.

Figure 8.10 Show/Hide button in the plug-in window header

Use this button to show or hide the plug-in window header while leaving the adjustable plug-in parameters visible.

Adjusting Plug-In Parameters

Once you have a plug-in inserted on a track with the plug-in window visible, you're ready to start adjusting the plug-in's parameters. This is where the fun begins! Logic Pro X provides a number of ways to adjust plug-in parameters, including the obvious (using the mouse), and the not-so-obvious (using the computer keyboard). Let's take a closer look.

Adjusting Plug-In Parameters with the Mouse

The most straightforward way to adjust plug-in parameters is to use the mouse. For vertical and horizontal slider controls, simply click and drag up/right to increase the value of the control or drag down/left to decrease the value. To adjust rotary knob plug-in parameter controls, click and drag vertically. To adjust numeric field parameters, click and drag vertically.

 If you click on a position just outside a rotary knob, the value will change to the clicked value. If you click and hold a value, the rotary knob value will follow with respect to the mouse's location.

You can modify the way that plug-in controls respond to the mouse using keyboard modifiers.

For fine adjustments:

- Hold **OPTION** while moving the control. (Hold **SHIFT** to invert this behavior for broader movements.)

To return a control to its default value:

- **OPTION-CLICK** on the control.

Adjusting Plug-In Parameters with the Computer Keyboard

It's actually a little-known fact that you can also use the computer's alphanumeric keyboard (or "QWERTY" keyboard) to adjust plug-in parameter controls in Logic Pro X.

To change the value of a plug-in parameter control incrementally using the computer keyboard:

1. Click in the text field for the control that you'd like to adjust.
2. Press the **UP ARROW** key to increase the value or press the **DOWN ARROW** key to decrease the value.

To set a value of plug-in parameter controls using the computer keyboard:

1. Double-click in the text field for the control that you'd like to adjust. (Once you're in a text field, you can use **TAB** or **SHIFT+TAB** to move to the next or previous text field, respectively.)
2. Type in a numeric value.

EQ Processing

EQ (short for equalization) is probably the most commonly used processor in the audio universe. In the consumer electronics world, EQ can be found just about anywhere that an audio signal is present; even the most humble car stereo usually has basic low, mid, and high EQ controls.

In the realm of audio production, EQ is an absolutely essential tool for getting individual instruments to sound their best, and for getting a multi-track mix to sound great.

Types of EQ

In the simplest terms, EQ is used to manipulate the frequency component of an audio signal. However, EQ processors themselves come in a variety of flavors that offer different numbers of EQ bands. These are often combined with different configurations within the available bands.

In the case of a simple graphic EQ, each band is represented by a slider control that is used to boost or cut the signal at that frequency. Graphic EQs use fixed frequency bands.

Figure 8.11 The Vintage Graphic EQ plug-in in Logic Pro X

By contrast, many of the EQs you will use in a DAW will have adjustable EQ bands of various types.

Some types of adjustable EQ bands include:

- **Parametric**—This is the most common type of EQ band. A parametric band is fully adjustable in terms of its frequency, gain, and Q (width) parameters.

- **High- and Low-Pass Filters**—These constitute another common type of EQ band, and are often abbreviated HPF and LPF, respectively. These filters permit signal to "pass" only above (HPF) or below (LPF) the selected frequency. High-pass filters are commonly used to eliminate low-frequency rumble, while low-pass filters can be employed to remove high-frequency hiss.

- **Shelf**—The shelf is similar to the high- and low-pass filter, but it is used to boost or cut the signal above (high-shelf) or below (low-shelf) the selected frequency by a specified amount.

- **Notch**—A notch filter is essentially a parametric band that has the gain setting fixed at the maximum negative value. This type of filter is used to "notch" out an offending frequency to make it inaudible.

Basic EQ Parameters

You'll find a number of standard EQ parameters on most software and hardware EQ processors. These will vary somewhat based on the type of EQ band.

Common EQ parameters include the following:

- **Frequency**—The Frequency parameter determines the target frequency for an EQ band. Every type of EQ band will have a frequency parameter, although some may be fixed at a specific frequency (such as the bands in a graphic EQ).

- **Gain**—The Gain parameter is used to boost or cut the volume of the associated EQ band. A gain control is typically found on parametric and shelf bands.

- **Q**—This is probably the most confusing EQ parameter for those who are new to audio production. The Q control is used to adjust the shape (slope or width) of an EQ curve. It behaves somewhat differently depending on the type of EQ band being used (parametric, shelf, filter).

- **Band On/Off**—This parameter can be used to enable or disable an EQ band, controlling whether that particular band is having an audible effect on the signal. This can be useful for doing an A/B comparison of the signal with and without an individual EQ band.

- **Input/Output** – The Input and Output gain controls can be used to boost or cut the signal at the input (before the EQ has been applied) or output (after the EQ has been applied).

EQ Plug-Ins in Logic Pro X

Logic Pro X offers a large selection of EQ plug-ins that cover a variety of production needs. Options include Channel EQ, Linear Phase EQ, Match EQ, and a Single Band EQ. Logic Pro X recently added a vintage EQ collection as well, featuring Vintage Console EQ, Vintage Graphic EQ, and Vintage Tube EQ. These plug-ins are modeled after three famous units from the 1950s, 1960s, and 1970s.

In addition, a number of other plug-ins feature adjustable EQ sections as part of the parameter settings.

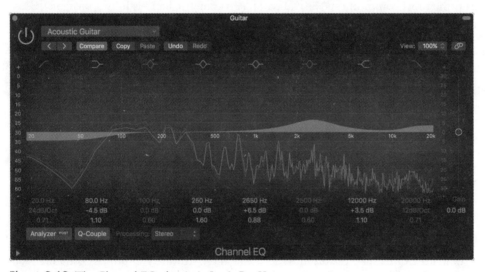

Figure 8.12 The Channel EQ plug-in in Logic Pro X

Strategies for Using EQ

Different types of material will require different approaches to EQ. But the basic techniques for finding good EQ settings are similar from track to track. Following are some tips for using Channel EQ and its 8 bands.

To use a Channel EQ on a track:

- Enable **BAND 1**, the high-pass filter, and increase the frequency enough to eliminate any unwanted noises at the extreme low frequencies, but not so much that the track loses significant low end.

On some tracks, such as drum overheads, you may want to eliminate almost all of the bass to make room for the kick drum track and other low-frequency instruments.

- Use the **LOW-MID**, **MID**, **HIGH-MID**, and/or **HIGH** bands to emphasize desirable qualities of a particular track.

   A common approach used to identify a target frequency is to boost the gain substantially (+12 dB or more) and slowly sweep the frequency control back and forth during playback. Listen carefully to the track while adjusting the frequency. Stop sweeping when you hear something you like, tone-wise. Then reduce the gain until the frequency is only slightly pronounced, compared to the original.

- Use the **LOW-MID**, **MID**, **HIGH-MID**, and/or **HIGH** bands as needed to reduce undesirable qualities of a particular track.

   A similar technique to the above can be used to identify target frequencies to cut. Simply stop sweeping when you hear something in the tone that sounds bad, intrusive, or unwanted. Then adjust the gain to a negative value that effectively reduces the prominence of the frequency, compared to the original.

- If necessary, use **BAND 2** (low shelf) or **BAND 7** (high shelf) to add low-end thump to a bass guitar or some high-end sizzle to cymbals (respectively).

- Use the plug-in's **MASTER GAIN** slider to increase or lower the output level since the resulting volume may have been affected by the EQ settings.

- Frequently **BYPASS** the plug-in to make sure that the changes you're making are actually improving the sound of the track and helping it to sit better in the overall mix!

> ⓘ You can load the Channel EQ plug-in by inserting one into an Audio Effect slot or by clicking the rectangular EQ display on the channel strip.

Dynamics Processing

Like EQ, dynamics processing is an essential part of the audio production workflow. Unlike EQ, dynamics processing is not particularly well known outside of audio production. It is fair to say that working with dynamics processors is a bit more complex than using EQ to boost the bass or cut the treble on your car stereo. But the basics of dynamics processing are pretty straightforward, and if you use your ears to dial in just the right settings you'll be working like a pro in no time!

Figure 8.13 The Vintage Opto Compressor in Logic Pro X

Types of Dynamics Processors

Dynamics processing refers to the manipulation of the volume (or amplitude) of an audio signal. Many instruments benefit from dynamics processing, including vocals, guitar, bass guitar, and drums. Depending on the situation, dynamics processors can be used to control a signal level that is varying dramatically, or to isolate just the loudest part of a signal while reducing or eliminating the quiet parts.

Some types of dynamics processors include:

- **Compressors**—The compressor is the most common type of dynamics processor. It is used to reduce the loudest part of an audio signal, resulting in a more predictable dynamic range. Once the loudest parts have been reigned in, it becomes possible to boost the entire signal without fear of clipping.

- **Limiters**—A limiter is essentially a compressor with a very high ratio setting (see the "Basic Dynamics Parameters" section below). As a result, the incoming audio signal is completely "capped" at a specified level that prevents clipping. Limiters are an essential part of the music production workflow and are often used to make an overall mix seem louder.

- **De-Essers**—A de-esser is a compressor that operates at a specific frequency. Its primary use is to reduce excessive "s" sounds in a recording, a phenomenon known as *sibilance*. This type of processor can also be used to eliminate other undesirable high-frequency sounds such as the breath noise in a flute recording.

- **Expanders**—An expander is a dynamics processor that can be used to reduce the level of a signal when it falls below a certain level. Expanders work something like a compressor in reverse. A gentle amount of expansion can help to reduce some of the ambient noise that gets picked up in an audio recording.

- **Gates**—A gate (often called a "noise gate") is a type of expander that will completely silence a signal when it falls below a certain level. Gates are frequently used to eliminate the "bleed" that occurs when a microphone picks up signal from an "off-mike" sound source, such as a snare drum being picked up by a kick drum mike. Gates are also useful to eliminate background noise in a recording, such as the buzz of a guitar amp that can be heard in "silent" parts of the performance. The purpose of a gate is to clean up the sound of a recording by helping to isolate the desired parts of the audio signal and eliminate the unwanted parts.

Figure 8.14 The Noise Gate in Logic Pro X

Basic Dynamics Parameters

Just like EQ processors, dynamic processors make use of a standard set of controls. These will vary somewhat based on the type of dynamics processor being used.

Common dynamics parameters include the following:

- **Threshold**—The threshold parameter sets the level that the input signal must reach in order to trigger the processor. Note that compressors and limiters are triggered when the signal exceeds the threshold, whereas expanders and gates are triggered when the signal drops below the threshold.

- **Ratio**—The ratio parameter determines the amount of signal reduction that occurs once the threshold has been reached. For example, when using a compressor with the ratio set to 2:1, the signal must exceed the threshold by 2 dB to result in an output increase of 1 dB. A limiter typically features a compression ratio of 100:1.

- **Knee**—The knee parameter determines how suddenly or gradually gain reduction is introduced in a compressor as the signal level approaches the threshold.

- **Gain**—The gain parameter is used to add a level increase (known as "makeup gain") to a signal that has been reduced in volume by a dynamics processor.

- **Attack/Release**—The attack parameter determines how quickly the dynamics processor will act once the audio signal has reached the threshold. The release parameter determines how quickly the dynamics processor will disengage once the threshold is no longer met.

Dynamics Plug-Ins in Logic Pro X

Logic Pro X offers nine types of dynamics plug-ins: the Adaptive Limiter, the Compressor, the DeEsser 2, the Enveloper, the Expander, the Limiter, the Multipressor, and Noise Gate. The basic functions of most of these are as described above. A few of the processors have specialized functions:

- The Enveloper allows you to shape the attack and release phases of a signal.

- The Multipressor, also known as a multiband compressor, divides a signal into multiple frequency bands allowing independent compression for each band.

Figure 8.15 The Multipressor in Logic Pro X

Applying dynamics processing to tracks can seem confusing at first. But, just like when working with EQ, there are some common strategies to help you find the right settings.

Strategies for Using Compression

To add compression to a track in Logic Pro X, you can add the Compressor plug-in to an Audio Effect slot on the target track's channel strip.

To use the Compressor on a signal:

1. Set the **Knee** to about 0.3, the **Attack** to about 10ms, and the **Release** to about 100ms.

2. Start off with a **Ratio** of about 3.0:1. (You may need a higher ratio if the track has a wide dynamic range.)

3. Reduce the **Threshold** setting until you start seeing a small amount of compression registering in the gain reduction (GR) meter. A good starting point is to target around 3 to 6 dB of gain reduction.

 You can switch between the Meter and Graph display to help visually monitor the effect of the compressor settings on the incoming signal level.

4. Adjust the **Make Up** gain control to increase the overall level of the track.

   Make sure that you don't increase the gain so much that the plug-in's output begins clipping. If you're having trouble finding the perfect gain setting, you may need to increase the ratio or lower the threshold.

 You can load the Compressor plug-in by inserting one into an Audio Effect slot or by clicking the Gain Reduction meter on the channel strip located just above the rectangular EQ display.

Strategies for Using a De-Esser

A de-esser is an essential tool when working with vocals. Even a great singer can have problems with sibilance in the studio. Fortunately, de-esser plug-ins have a small number of controls that you can easily learn to use effectively.

 For best results, insert the De-Esser 2 after any compressor plug-ins on the track.

To use a De-Esser 2 on a vocal track:

1. Select the **Highpass Shelving** button of the Filter section to limit the effect to the high frequencies.

2. Turn on **Filter Solo** and play back a sibilant portion of the track.

 The Filter Solo button lets you isolate the sibilance targeted by the Frequency control. This lets you hear the sibilant peaks that will be reduced by the plug-in.

3. Adjust the **Frequency** control while listening carefully; try to zero in on just the harsh sibilant frequencies.

> ⓘ Sibilant sounds typically occur in the 4 to 10 kHz range. Most singers will require de-essing in the 6 to 7 kHz range.

4. Once you've found the target frequency, stop playback and turn off **FILTER SOLO**.

5. Set the **THRESHOLD** control to -0.0 dB and begin playback across the sibilant section again. Watch the gain reduction (GR) meter to help identify when de-essing is being applied.

6. Slowly lower the Threshold value, as needed, to set the level at which de-essing is triggered.

   - While setting the Threshold, toggle the **BYPASS** control frequently to compare the result to the original vocal.

   - Unsolo the track from time to time to hear how the de-essed vocal sounds in the overall mix. You don't want to overdo things; a little de-essing can go a long way!

Reverb and Delay Effects

Reverb and delay effects are indispensable tools for creating a sense of space. Collectively known as "time-based effects," reverb and delay can be used to create a wide range of sounds. This range spans the gamut from providing tasteful spatialization of audio, to adding a subtle spaciousness to a track, to creating dramatic effects that completely alter the sound of a voice or instrument.

What Is Reverb?

Reverb effects are used to create the sound of a real or imaginary space. Reverb algorithms can be used to recreate the sound of small spaces (rooms, cars, phone booths) or large spaces (concert halls, parking garages). A number of types of reverbs are available in both hardware and software formats. Some use a series of delays (see "What Is Delay?" below) to create smooth, musical-sounding results. Other reverbs use a technique called "convolution," which actually uses samples of real-world acoustic spaces to create incredible realism.

Figure 8.16 The Space Designer plug-in in Logic Pro X

Reverb in Logic Pro X

Logic Pro X offers a number of reverb plug-ins, ranging from basic to advanced. Reverb plug-ins include ChromaVerb, EnVerb, SilverVerb, and Space Designer. The first three are digital reverb effects that include additional options. The Space Designer is a convolution reverb effect that utilizes recorded and synthesized impulse responses. This allows for incredibly realistic recreations of real-world spaces as well as more adventurous sounds.

Common reverb parameters include the following:

- **Decay**—The decay control, sometimes referred to as Reverb Time, determines how long it will take for the reverb to "fade out" after a sound has been processed.

- **Pre-Delay**—The pre-delay control adjusts the amount of time that a reverb "waits" after receiving an input signal before it begins to create reverb.

- **Diffusion**—When set to a high value, the diffusion control will emphasize the initial build-up of echoes. Low values will reduce the initial build-up, which can result in better clarity.

- **Low-Pass and High-Pass Filter**—Just like the filter controls on an EQ, this control can be used to eliminate content above or below the specified frequency.

- **Room Type/Size**—This determines the type or size of space being emulated. Depending on the plug-in, this can correlate to reverb time (based on size of the space) or diffusion (based on the room materials). In some cases, selecting a room type or size may adjust a number of other reverb parameters.

- **Mix**—The Mix control can be used to adjust the wet/dry balance of the plug-in's output. (See "Wet Versus Dry Signals" below.)

Applications for Reverb Processors

There are many creative ways to use reverb in a project. Whether used to add a subtle sense of space to a vocal or to completely wash out a guitar for an ambient bed, reverb can truly bring a track to life.

Suggestions and scenarios for reverb usage:

- Add some character to a snare track by inserting a small room reverb directly on the track. Then adjust the wet/dry control to get the proper balance. Try the **ROOM** preset in SilverVerb.

- Give a vocal track a vintage sound by using a medium plate reverb. Try the **SMALL SPACES > PLATE REVERBS > 1.1S VOCAL PLATE** preset using Space Designer.

- Create a distant, washed-out guitar sound by using a large church reverb and setting the wet/dry control to 100% wet. Try the **LARGE SPACES > INDOOR SPACES > 08.7S BIG GOTHIC CHURCH** preset using Space Designer.

- Add a medium hall reverb in a send-and-return configuration to apply reverb to a number of tracks in a project including vocals, acoustic guitars, keyboards, and more. This can help "gel" the mix by putting all of the components into the same acoustic space. Try the **HALLS > VINTAGE CONCERT HALL** preset using ChromaVerb.

What Is Delay?

Delay is the term that audio engineers use for a processor that creates an echo. Delay is a simple effect in concept, but over the years both hardware and software delays have evolved into extremely complex processors.

Delay plug-ins provide dozens of different approaches to delay, from digital models of vintage tape-style echos to modern granular pitch-shifted delays. And almost all modern delays have the ability to synchronize their individual echoes to the tempo of a song.

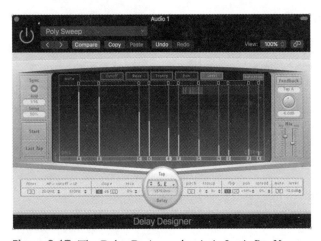

Figure 8.17 The Delay Designer plug-in in Logic Pro X

Delay in Logic Pro X

Logic Pro X offers a number of delay plug-ins. The Delay Designer is a multitap delay with additional control over level, panning, pitch, and filters for each tap, making it one of the most robust delay options available. The Echo plug-in is a simple delay with minimal features. The Sample Delay is a utility that allows you to delay the left and right channels in very fine increments (measured in samples or milliseconds). The Stereo Delay lets you independently set delay time, feedback, and other parameters per channel. Finally, the Tape Delay emulates vintage tape echo machines and provides subtle saturation to the signal.

Common delay controls include the following:

- **Delay Time**—Set the plug-in delay time in milliseconds or musical values linked with the project grid when Sync is enabled. Ranges can vary by plug-in.

- **Feedback**—The feedback control determines how much of the plug-in's output will be "fed back" into the input. The feedback setting ultimately controls the number of echoes that will occur.

- **Filters**—Like on an EQ, some delay plug-ins have filter controls that can be used to eliminate content above or below the specified frequency. These can include: low-pass, high-pass, low cut, high cut, and other filtering options.

- **Sync**—When enabled, the Sync control will cause the delay time to synchronize to the session tempo.

- **Mix**—The Mix control can be used to adjust the wet/dry balance of the plug-in's output. (See "Wet Versus Dry Signals" below.)

Applications for Delay Processors

A little delay can help add excitement to your mix. This is a great way to propel a track forward and add rhythmic variety.

Suggestions and scenarios for delay usage:

- Use Selection-Based Processing to quickly apply delay to just the last word of a vocal phrase.

  - With the Marquee tool, select the target word or phrase.

  - Open the Selection-Based Processing window. (In the Tracks area of the main window, select **FUNCTIONS > SELECTION-BASED PROCESSING** or press **OPTION+SHIFT+P**.)

  - Under the blue **Set A** column, add the desired plug-in and configure the parameters.

  - Audition the processing by clicking the **PREVIEW** button (speaker icon) in the lower left corner of the plug-in window and experiment with the plug-in's wet/dry control to get the proper balance.

  - Check the boxes for **SPLIT AT MARQUEE BORDERS** and **ADD EFFECT TAIL**. (This ensures that the effect doesn't cut off abruptly during playback.)

  - Apply the Selection-Based Processing by clicking the **APPLY** button.

- Add some width to a mono acoustic guitar track by applying the Sample Delay plug-in with delay on the left or right channels. Alternatively, apply slightly different delay times to the left and right channels with the Stereo Delay plug-in.

- Use the Echo or Tape Delay to apply a simple quarter-note delay in a send-and-return configuration to add sustain, rhythmic emphasis, and interest to a variety of track types including vocals and guitar solos.

- Get an Edge-like guitar sound (U2) by running a simple guitar part into a dotted-eighth-note delay synced to the song's tempo. Set the feedback parameter high enough to give at least 2 to 3 repeats of each note. Take it a step further with the Stereo Delay plug-in by setting both channels to different rhythmic values to create an even more complex rhythmic effect.

The Pedalboard plug-in located in the Amps and Pedals folder of the plug-in menu has a large assortment of effect pedals including multiple delay options. While the Pedalboard was designed with guitarists in mind, these can easily be used with any audio or software instrument source.

Wet Versus Dry Signals

When working with time-based effects such as reverb and delay, you will need to balance the original signal against the processed signal. Audio producers refer to the original signal as the "dry" signal and the reverberated or delayed signal as the "wet" signal. Here are some common ways to adjust the wet/dry balance.

Using Time-Based Effects as Plug-in Inserts

Although time-based effects are typically applied using a send-and-return configuration (see below), at times you may use a reverb or delay as an Audio Effect directly on the source track. For example, you may want to add reverb to a snare track to give it a unique sound, or you may want to create a swirling guitar wash, where the signal is 100% wet. In these scenarios, you can place the desired plug-in directly on the target track's channel strip.

When applying a time-based effect directly to a track, the only way to adjust the wet/dry balance is to use the plug-in's controls. The location and labeling of these controls can vary from plug-in to plug-in. Most of the delay plug-ins in Logic Pro X use wet/dry sliders to adjust the mix.

Figure 8.18 Tape Delay's wet/dry balance is adjusted using two sliders in the Output control area.

The Output control of the Tape Delay, for example, uses dry and wet sliders to control the overall output of the corresponding signal. You can also type directly into the number field at the top of the slider and manually enter a numerical value from 0% to 100%.

 As a general rule, you'll want to set the plug-in to a relatively dry setting when emulating "natural" reverb or delay characteristics. Start with a low setting, such as 70% Dry and 30% Wet, as shown in the example in Figure 8.18, and adjust from there as needed.

Using Send-and-Return Configurations

Configuring your time-based effects using a send and return is typically preferable, for a variety of reasons. First, you'll often want to apply the same reverb settings to many tracks in your project, so using a single plug-in is much faster to set up and adjust.

Second, because time-based effects can use significant CPU power, using a send and return will save on CPU processing. You can create one instance of a reverb or delay on an Aux Input track and then use as many sends to the track as necessary to apply the effect for multiple tracks.

And third, each source track's direct output will remain completely dry, while the send level can be used to adjust the wet signal. This provides a quick and easy way to modify the wet/dry balance for a whole bunch of tracks in your project without needing to change the settings of the plug-in itself.

Although advanced send-and-return configurations are beyond the scope of this book, you can configure a basic setup with a few simple steps. Use the techniques outlined below to get started with a reverb send.

 The same techniques can be used to create a delay send. Simply modify the name you use for the aux channel strip and change the plug-in you assign to the channel.

To set up a reverb send across multiple tracks:

1. In the Mixer, select all of the tracks that you'd like to send to the reverb by clicking on their channel strips.
2. Click on an empty Send slot on one of the selected channel strips.
3. Select an available option from the list of busses. (For example, choose **Bus 1**.) A send to the selected bus will be created across all of the selected channel strips, and an aux channel strip will be created with the selected bus routed to its input slot.

To set up the reverb return:

1. Rename the new aux channel strip to something meaningful, like **Reverb**.
2. Assign your preferred reverb plug-in on one of the Audio Effect slots on the aux channel strip.

To set the send levels:

- For each of the source channel strips you used in Step 1, bring up the Send Level knob as needed to introduce reverb for that track.

Review/Discussion Questions

1. How can you insert a plug-in on a track? How can you remove an inserted plug-in? (See "Inserting Plug-Ins on Tracks" beginning on page 190.)

2. How can you move a plug-in from one track to another track? How can you duplicate a plug-in? (See "Moving and Duplicating Plug-Ins" beginning on page 192.)

3. What is the purpose of the previous (<) and next (>) buttons in the plug-in controls? (See "Plug-in Settings Controls" beginning on page 193.)

4. Is it possible to change the plug-in that you are viewing in the currently open plug-in window? How might you enable/disable this behavior? (See "Link Control" beginning on page 194.)

5. What is the function of the Multi setting for the Link control? (See "Link Control" beginning on page 194.)

6. Which keyboard modifier can you use to make fine adjustments on plug-in controls? (See "Adjusting Plug-In Parameters with the Mouse" beginning on page 195.)

7. Which keyboard modifier can you use to reset a plug-in control to its default value? (See "Adjusting Plug-In Parameters with the Mouse" beginning on page 195.)

8. In what ways can plug-in parameters be adjusted using the computer keyboard? (See "Adjusting Plug-In Parameters with the Computer Keyboard" beginning on page 196.)

9. What are the four common types of EQ bands? (See "Types of EQ" beginning on page 196.)

10. What EQ plug-in options are included with Logic Pro X? (See "EQ Plug-Ins in Logic Pro X" beginning on page 198.)

11. What dynamics plug-ins are found in Logic Pro X? (See "Dynamics Plug-Ins in Logic Pro X" beginning on page 202.)

12. What is the difference between reverb and delay? What category of effects do both of these processes belong to? (See "Reverb and Delay Effects" beginning on page 204.)

13. What are some scenarios in which you might place a time-based effect directly on a track? (See "Using Time-Based Effects as Plug-in Inserts" beginning on page 208.)

14. What are some advantages of using a send-and-return configuration for time-based effects? (See "Using Send-and-Return Configurations" beginning on page 209.)

To review additional material from this chapter and prepare for certification, see the Logic Audio Production Basics Study Guide module available through the Elements|ED online learning platform at ElementsED.com.

Exercise 8

Optimizing Tracks with Signal Processing

🎧 Activity

In this exercise, you will learn how to use plug-ins to optimize tracks in Logic Pro X. You'll begin by looking at some practical applications for equalization. Then, you'll explore some options for controlling dynamics. Finally, you'll create a send-and-return configuration to apply reverb for multiple tracks.

⏱ Duration

This exercise should take approximately 20 minutes to complete.

⊕ Goals/Targets

- Use the Channel EQ plug-in to cut bass frequencies and boost high frequencies
- Use the Compressor plug-in to reduce variations in peak volume levels
- Set up a send to an aux channel strip
- Apply reverb processing using the Space Designer plug-in
- Save your work for use in Exercise 9

Exercise Media

This exercise uses media files taken from the song, "Lights," provided courtesy of Bay Area band Fotograf.

Written by: Zack Vieira and Eric Kuehnl; Performed by: Fotograf

The media provided for this course may be used for educational purposes only. No rights are granted to use the media for any other personal, commercial, or non-commercial purposes.

Getting Started

To get started, you will open your completed project from Exercise 7. This will serve as the starting point for this exercise.

Open your existing Lights project:

1. Launch Logic Pro X and select **FILE > OPEN** or press **COMMAND+O**. The Open dialog box will display.

2. Navigate to the location where you saved your project in Exercise 7.

3. Select your Lights02-xxx project and click **OPEN**. The project will open as it was when last saved.

4. Choose **FILE > SAVE AS** and create a copy of the project called Lights03-xxx keeping the project in the original project folder.

Applying EQ and Dynamics to Tracks

To get an idea of where EQ processing might be useful, you should take a moment to listen to the project in its current state. Pay particular attention to any areas where frequencies are clashing between multiple tracks. Also listen for tracks that could benefit from more controlled dynamics.

Evaluate the tracks:

1. Press the **SPACEBAR** to begin playback.

2. Listen for tracks that compete for space in the same frequency range, such as the 01 Kick track and the 05 Synth Bass track. Try to formulate some ideas about how to give each track its own space in the frequency spectrum.

3. Also, listen for tracks with dynamics that might cause problems. Listen for consistency in the volume of notes for tracks that should serve as a steady backbone, such as the bass guitar during the choruses.

4. Feel free to solo and mute tracks to isolate certain elements. Adjust the faders to see if you can find a good balance of tracks. See if certain tracks present unique problems.

   - Is it hard to separate certain tracks from one another without causing one to obscure the other in a given frequency range?

   - Is it hard to find a level for certain tracks that allows all of the notes to be heard without causing certain notes to get too loud?

Adding EQ Processing

When using EQ, you may find yourself wanting to boost frequencies to emphasize the best parts of a track. However, it is equally important to cut frequencies to create room for other tracks to be heard. In this section of the exercise, you will use EQ to reduce the bass frequencies in the 05 Synth Bass track while slightly boosting the high frequencies in order to separate it from competing bass frequencies in the 01 Kick track.

Insert an EQ plug-in on the 05 Synth Bass track:

1. Locate the Audio Effect section on the 05 Synth Bass channel strip.

2. Click on the first Audio Effect slot position and select **EQ > CHANNEL EQ > STEREO** (or click on the rectangular EQ display just above the input section of the channel strip). The Channel EQ plug-in interface will display.

Adjust the EQ settings:

1. Turn on the EQ bands that you expect to use. The default settings have all bands enabled except for the High-Pass and Low-Pass Filters (**Band 1** and **Band 8**, respectively).

    - Enable Band 1 by clicking on the **BAND 1 ON/OFF** button so that it becomes lit in red.

Figure 8.19 Clicking on the Band 1 On/Off button to enable the High-Pass Filter band (shown active)

2. Set the parameters for each EQ band:

    - **Band 1 (High-Pass):** set the Q to 24 dB/oct (maximum) and the Frequency around 65.0 Hz.
    - **Band 2 (Low Shelf):** set the Q to 1.00 (default), the Frequency to 200.0 Hz, and the Gain to around -3.0 dB.
    - **Band 5:** set the Q to 1.20, the Frequency to 800.0 Hz, and the Gain to around -2.0 dB.
    - **Band 7 (High Shelf):** set the Q to 0.50, the Frequency to 5.00 kHz, and the Gain to 2.0 dB.

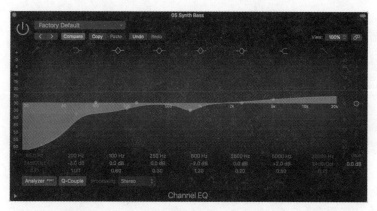

Figure 8.20 The Channel EQ settings for the 05 Synth Bass track

3. Close the Channel EQ plug-in window when finished.

Adding Dynamics Processing

Next you'll apply dynamics processing to help control the signal levels. The **08 Bass Gtr** track has a slightly wider dynamic range than ideal, making it hard to position in the mix without constantly adjusting the volume fader. Here you will rein in the dynamic variation using a compressor.

 A great way to learn to use plug-ins such as EQ and Dynamics is to start from a relevant factory preset.

Insert a Compressor on the 08 Bass Gtr track:

1. Locate the Audio Effect section on the **08 Bass Gtr** track.

2. Click on the first slot and select **DYNAMICS > COMPRESSOR > STEREO** (or click the Gain Reduction meter located above the rectangular EQ display on the channel strip). The Compressor plug-in interface will display.

Use a factory preset to compress the 08 Bass Gtr track:

1. From the Settings menu, choose the **03 GUITARS > FET BASS** preset.

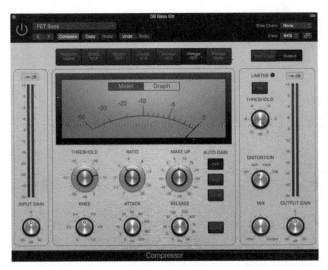

Figure 8.21 The Compressor settings for the FET Bass preset

2. Position the playhead at Bar 48, just before the first chorus.

3. Play through the first chorus of the song and listen to the **08 Bass Gtr** track. If you don't see movement on the compressor's VU meter, try lowering the Threshhold until you see the meter consistently bouncing around -5 dB. (Try around **-20 dB** as a starting point.)

   The dynamic variation should be dramatically reduced.

4. Adjust the **MAKE UP** control on the Compressor to find a good volume level for the track. (Try around **10.0 dB** as a starting point.)

5. Close the Compressor plug-in window when finished.

Using a Send-And-Return Configuration for Reverb

In this section, you'll add some reverb to help smooth out the vocals and give them some acoustic space. You'll set up a send-and-return configuration that can be used for multiple tracks in the project.

Send audio from the vocal track to an aux channel strip:

1. Locate the Sends section on the **06 Vox** track.

2. Click on the first Send slot and select an available Bus, for example: **BUS > BUS 4**. A send to Bus 4 will be created, and an aux channel strip will be created with the Bus routed to its input.

Recording engineers commonly put effects returns after instrument tracks and just before the Master channel or Stereo Out for the purposes of signal flow. Logic Pro X adopts this practice when positioning the aux channel in the Mixer.

Assign a reverb plug-in to the aux channel strip:

1. Choose **VIEW > SHOW MIXER** or press **X** to open the Mixer tab.
2. Locate the Audio Effect section on the newly created **AUX** channel strip.
3. Click on the first slot and select **REVERB > CHROMAVERB > STEREO)**. The ChromaVerb plug-in interface will display.
4. From the Settings menu, choose the **SPACES > EMPTY HALL** preset.
5. Double check that the Dry slider is set to 0% and the Wet slider is set to 100%.

Figure 8.22 The ChromaVerb plug-in with the suggested settings

6. Close the plug-in window when finished.
7. Rename the aux channel strip as **Reverb**.
8. As a final step, **CONTROL-CLICK** on the **Reverb** channel strip's Solo button to enable solo-safe mode. This will prevent the track from muting when a source track is soloed.

Send audio from the 06 Vox track to the Reverb channel:

1. Return to the Send slot containing Bus 4 on the **06 Vox** channel strip.
2. Increase the level of the Send Level knob to about -3 dB. You should now be able to hear the reverberated vocals when you play the project.

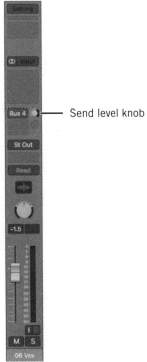

Figure 8.23 The send controls for the reverb send on the Smart Strings track

Finishing Up

To complete this exercise, you will need to save your work and close the project. You will be reusing this project in Exercise 9, so it's important to save the work you've done.

Before you wrap up, you can also listen to the project to hear the processing changes you've made.

Finish your work:

1. Choose **FILE > SAVE** (or press **COMMAND+S**) to save the project.
2. Press **RETURN** to position the playhead at the beginning of the project.
3. (Optional) Press the **SPACEBAR** to begin playback and listen through the project.
4. Press the **SPACEBAR** a second time when finished to stop playback.
5. Choose **FILE > CLOSE PROJECT** to close the project.

That completes this exercise.

Chapter 9

Finishing a Project

...*What You Need to Do to Create a Stereo Mixdown*...

This chapter introduces you to the processes of adding automation and creating a stereo mixdown of your Logic Pro X project. We cover the five automation modes in Logic Pro X, what their differences are, and how to use them. We also cover a variety of editing techniques that you can use to create or fine-tune the automation on your tracks. We then cover uses for limiters and dither plug-ins on the master bus before wrapping up with a discussion on exporting your mix as a stereo file.

⊕ Learning Targets for This Chapter

- Understand what settings will and will not be recalled with a project
- Learn how to select the appropriate automation mode for a desired result
- Learn how to write automation changes in Touch and Latch mode
- Understand how to display and edit automation graphs
- Add appropriate processing to a Stereo Output
- Use the Export command to create a stereo mix of a project

Key topics from this chapter are illustrated in the Logic Audio Production Basics Study Guide module available through the Elements|ED online learning platform. Sign up at ElementsED.com.

After you have finished balancing your tracks, adding processing to the tracks, and adding any desired effects for the mix, it's time to start thinking about finishing up the project and creating a stereo mixdown. As you put the finishing touches on your mix, you can use automation to create and save dynamic changes. Then you can export your mix as a stereo file, configuring the file parameters as needed.

Recalling a Saved Mix

When you open a project that you've previously worked on, all of the Logic Pro X mixer settings are recalled with the file, including the following:

- **Track layouts and views**—The track order, track heights, and zoom settings that were in place when the project was last saved will all be recalled.

- **Mix attributes, settings, and automation**—The settings for volume faders and pan controls will be recalled for each track, along with any automation you previously created.

- **Track and send routing**—Any routing you've used for submixes, effects sends, and similar configurations will be recalled, along with the send levels and send pan settings.

- **Plug-in assignments**—Plug-in assignments and parameter settings will be recalled for each track.

Occasionally, however, conditions may exist that will prevent a project from recalling exactly as it was when last saved. These conditions can include the following:

- **External effects processors**—As discussed in Chapter 7, if you've used external gear for your mix, you will need to reconnect the gear and reconfigure all the hardware settings whenever you return to the project.

- **Unavailable plug-ins**—If you open the project from a system that does not have the same plug-ins installed as the system where you created the project, any unavailable plug-ins in the project will gray out on screen and will not provide processing.

- **Unavailable media**—Any audio that might be missing from the Project's folder will not be available for playback.

Using Automation

On a simple project, you may be able to set the volume faders, pan controls, and other parameters on tracks to a static position and leave them unchanged from the start of the mix to the end. But often a mix will require dynamic changes during the course of playback. This is where track and region automation can be useful.

Logic Pro X allows you to record changes you make during playback as automation on your tracks or regions. Automatable parameters include track volume, pan, and mute controls; send controls; and plug-in controls.

Selecting an Automation Mode

Logic Pro X provides six automation modes. The currently active automation mode on each track determines how automation will function for the track. You can set the automation mode for each track independently, using the Automation Mode selector for a channel strip in the Mixer. You can also select the automation mode in the Main window once you enable the Show Automation function, as discussed below.

Figure 9.1 Automation Mode selector in the Mixer window

Available Automation Modes

The automation modes available in Logic Pro X include the following:

- **Read mode** (default)—This mode plays back existing automation on the track but will not write any new automation. Use this mode when you want a track to play back automation you have previously recorded for the track, but you do not want to record new changes as automation.

- **Latch mode**—This mode writes automation only on parameters that you change during the automation pass. Use this mode to write automation selectively for some track settings without locking in other settings.

- **Touch mode**—This mode writes automation only on parameters that you change, and only while you are actively modifying those parameters. Use this mode to change certain track settings only in specific areas of the track.

- **Write mode**—This mode writes automation for all parameters whenever the transport is rolling. Use this mode if you want to write real-time automation on a track and lock in the volume, pan, send, and plug-in settings.

- **Trim mode**—This mode offsets the value of existing automation (Volume, Pan, Send level) by adjusting it up or down by the amount that you move the fader or a control. This mode preserves existing dynamic changes, offsetting them by the trim amount. Trim is a secondary mode that works in combination with Touch/Latch modes.

- **Relative mode**—This mode adds an independent, secondary automation curve which offsets the original primary curve for a selected parameter. Both the initial primary curve and the new secondary curve will be visible and can be edited. Relative is a secondary mode that works in combination with Touch, Latch, and Write automation modes.

Enabling the Show Automation Function in the Main Window

To display current automation modes and see automation graphs in the main window you have to enable the Show Automation button just above the track list. Alternatively you can press the **A** key to activate the Show Automation function.

Figure 9.2 Enabling the Show Automation button

Selecting an Automation Mode

You can select an automation mode for a track or channel in either the main window or the Mixer window.

To select an automation mode, do the following:

1. Click on the Automation Mode selector in the Mixer or Main window.

2. Select the desired mode from the pop-up menu. (See Figure 9.3.)

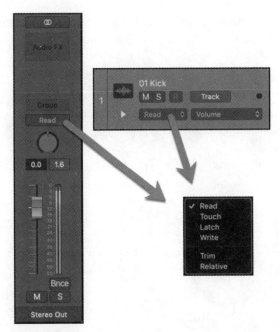

Figure 9.3 Selecting an automation mode: channel strip in the Mixer window (left) and Track List in the Main window (right)

Writing Real-Time Automation

When writing real-time automation, you will commonly use Latch mode for basic settings on your initial automation passes. This can be useful when writing automation across long sections.

 By way of example, you might use Latch mode to automate changes for different sections of a song, such as to increase the level of the lead vocals during the choruses and return to a lower level during the verses.

To automate basic settings, do the following:

1. Put the target track(s) into Latch mode.
2. Begin playback. Automation will not be writing at this point.
3. When you reach the location where you need to make a change, adjust the volume, pan, or other settings on the target track(s). The changes will begin writing as new automation and will continue until you make the next change.
4. Continue playback through the project, adjusting the settings during each section as needed.
5. Return the track(s) to Read mode when finished (for playback purposes).

With your initial track settings and automation in place, you can use Touch mode to refine the mix in specific areas. This can be useful to touch up existing automation or to make dynamic changes that affect only small areas of the project.

 By way of example, you can use Touch mode to increase the send levels for reverb or delay at the ends of certain vocal lines, while automatically returning to previous levels in between the changes.

To add or modify automation in specific areas, do the following:

1. Put the target track(s) into Touch mode.

2. Begin playback. Automation will not be writing at this point.

3. When you reach the location where you need to make a change, adjust the volume, pan, or other settings on the target track(s). New automation will be written and will continue while you hold the control in place.

4. Release the control to return the parameter to its previous setting.

5. Continue playback through the project, making additional changes in selected areas as needed.

6. Return the track(s) to Read mode when finished (for playback purposes).

Viewing Automation Curves

Each automatable parameter on a track has an associated automation graph, or automation curve as they are known in Logic. The automation curve can be displayed in the Edit window, allowing you to see a visual representation of the automation changes.

To display an automation curve, do the following:

1. Click the **SHOW AUTOMATION** button to enable it, or press the **A** key. (See Figure 9.2 earlier in this chapter.)

2. Select the type of automation that you want to display from the Automation Parameter pop-up menu. (See Figure 9.4.) The Automation Parameter will be set to Volume by default. The automation curve will be displayed as a faint orange line superimposed on the track. Clicking onto any region to select it will make the line more prominent.

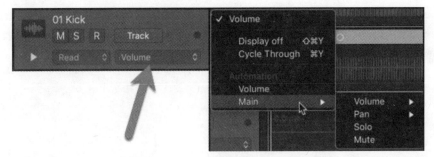

Figure 9.4 Selecting the Volume automation view (left); the Volume automation graph displayed (right)

Figure 9.5 The volume automation curve displayed on the 01 Kick track

As an alternative, you can display an automation curve in an automation subtrack beneath a target track using the **Subtrack Disclosure Triangle** at the head of a track. On Audio tracks, the Volume automation lane will appear by default. Additional automation lanes can be displayed by clicking on the plus sign (+) that appears when you place the pointer over the header of a displayed subtrack lane.

Figure 9.6 Clicking the Subtrack Disclosure Triangle button (left); Volume automation subtrack displayed (right)

Editing Automation Points

With an automation curve displayed, you can edit the curve using its automation points. Automation points are locations on the automation graph where the automation line changes slope or direction.

Using the Pointer Tool for Automation

You can edit the automation graph by adding, moving, or deleting automation points using the Pointer tool. To edit automation with the Pointer tool, do any of the following:

- Click on the automation curve line to add an automation point.
- Click and drag an existing automation point to adjust its position.
- **Double-click** on an existing automation point to remove it.

Figure 9.7 Editing the Volume automation using the Pointer tool

Automation can also be edited using other tools, such as the Pencil tool.

Using the Pencil Tool for Automation

The Pencil Tool can be useful for drawing automation changes on an automation curve. The Pencil Tool can be combined with Snap Automation to create a variety of automation shapes.

To draw a freehand automation shape, do the following:

1. In the main window press the **T** key to bring up the left-click tool menu.

2. Verify that the **PENCIL TOOL** is selected.

Figure 9.8 Selecting the Pencil Tool

3. Click and drag on an automation curve to draw a new automation shape. A series of new automation points will be created, matching the shape you draw.

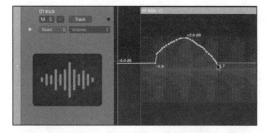

Figure 9.9 Drawing freeform automation with the Pencil tool

You can also create stepped automation with the Pencil Tool by activating the Snap Automation function. In this mode, the Pencil Tool will create automation in short line steps rather than curves as you click-drag. You can configure the length of each step in the Snap pop-up menu in the main window (See Figures 9.10 and 9.11.)

Figure 9.10 The Snap pop-up menu next to the blue power button

To create stepped automation, do the following:

1. Configure a rhythmic value for the automation grid by clicking the Snap pop-up menu, selecting **Snap Automation**, and choosing a rhythmic value. This is set to **Automatic** by default, meaning the snap increment will change based on your zoom level.

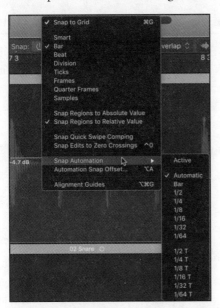

Figure 9.11 Selecting a rhythmic value for stepped automation in the Snap pop-up menu under the control bar

> When active, the Snap Automation function causes automation edits to snap to the chosen grid value.

2. With the Pencil Tool selected, hold the **Option** modifier key.

3. Click and drag on an automation graph to draw a new stepped segment. Stepped automation can be useful for writing changing settings of an exact length, such as rhythmic changes to a plug-in parameter, like filter cutoff on a synth. In Figure 9.12, panning automation has been drawn on a 1/4 note grid, alternating from side to side.

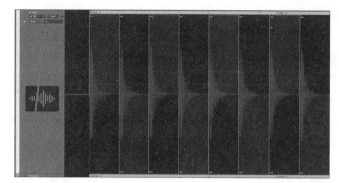

Figure 9.12 Drawing a stepped automation segment with the Pencil Tool

4. Release the mouse at the point where you want the steps to end.

Creating a Mixdown

Mixing down is the process of recording the output from multiple tracks to a stereo file. This process is also commonly referred to as *bouncing* the project. Mixing down is often the last phase of audio production, although you can create a bounce at any time to create a complete mix as a stereo file.

Adding Processing on the Stereo Output

Prior to creating a final mix, you may want to consider using some final dynamics processing on the entire mix to optimize your output levels. You can also add a dither plug-in to preserve the low-level dynamics in your mix during a bit-depth reduction.

Adding a Limiter to Optimize Output Levels

As discussed in Chapter 8, a limiter is a dynamics processor that caps the audio signal at a specified level. A limiter can be used on the Stereo Out channel to protect against clipping. Using a limiter can also help make the overall mix seem louder.

To use a limiter on your mix in Logic Pro X, do the following:

1. Click on an Audio FX insert position on the Stereo Out channel and select **DYNAMICS > LIMITER > STEREO**. (See Figure 9.13.)

Finishing a Project   **229**

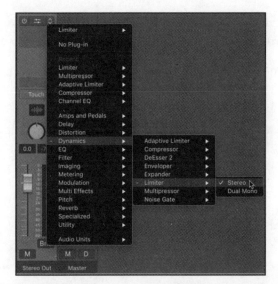

Figure 9.13 Selecting a limiter plug-in for the Stereo Out track

The Limiter plug-in window will open.

Figure 9.14 The Limiter plug-in window includes Gain, Release, Output Level, and Lookahead controls.

2. Play through the project, taking note of the peak readings displayed on the **INPUT** meter.

3. Adjust the **GAIN** control up or down until you have a strong input signal level that stays between −10 and 0 on the meter through most of the song. The signal should occasionally trigger the gain reduction meter (labeled **Reduction**) during the loudest peaks.

> You generally want to see gain reduction of no more than −3 to −6 dB in the loudest areas of the song.

4. Adjust the **Output Level** control to set the maximum signal level going to your audio interface. To prevent clipping in your mixed file, set this control to a negative value. Try around –0.5 to –1.5 dB.

5. Set the **Release** control to its minimum value of 2.0 ms to provide a quick response for the momentary peaks you are targeting with this type of limiter. Adjust this value upward during playback as needed to prevent sudden, unnatural-sounding volume changes.

6. Leave the **Lookahead** control set to its default value of 5.0 ms. This feature scans ahead to prepare the limiter for any potential upcoming peaks.

 A different option you can use on the Stereo Out channel is the Adaptive Limiter. This processor functions similar to the Limiter but is designed specifically to maximize the output level of a mix.

Creating a CD-Ready Mix

To create a CD-ready mixdown from Logic Pro X, you will need to perform a bit-depth reduction: your mixdown file will need to use 16-bit audio in order to be burned onto a CD, whereas Logic Pro X uses 24-bit audio files by default.

 The Red Book audio CD standard requires audio files that are encoded with a sample rate of 44.1 kHz and a word length of 16 bits.

As discussed in Chapter 3, lower bit-depth audio also exhibits a reduced dynamic range. To help preserve the dynamic range when reducing bit depth, you can add *dither* to your project. Dither is a form of randomized noise used to minimize signal loss when audio is near the low end of its dynamic range, such as during a quiet passage or fade-out.

Adding Dither to a Mix

Proper use of dithering allows you to squeeze better subjective performance out of 16-bit or 8-bit audio.

In Logic Pro, dithering is applied during the bounce process. To create a mix for CD, you would select the dither checkbox during the final bounce.

 If you will be keeping your mixed file at 24-bit resolution, you should NOT use dither. Doing so will needlessly add noise to the mix.

To add dither to your mix during a bounce, do the following:

1. Select **File > Bounce > Project or Selection**. (See Figure 9.15.)

Finishing a Project   231

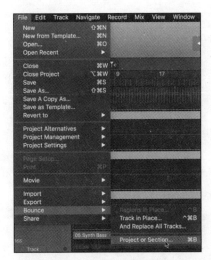

Figure 9.15 Selecting the Bounce option from the file menu

The Bounce dialog box will display.

2. Choose the type of dither noise shaping to use for your mix.

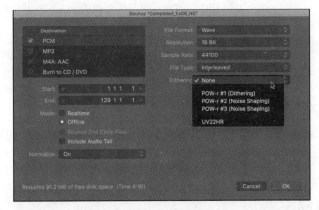

Figure 9.16 The bounce project window showing the Dithering options menu

Dither Noise Shaping Options

Noise shaping is a method of improving the signal-to-noise ratio of dither. POW-r Dither noise shaping improves audio performance and reduces the perceptible noise in the dither by shifting it into a less audible range.

POW-r #1: A dithering curve with a flat frequency spectrum is utilized to minimize quantization noise. This option is recommended for solo instrument recordings and spoken-word recordings.

> **POW-r #2:** Additional noise shaping is used over a wide frequency range. This can extend the dynamic range of the bounce file by 5-10db. This is suitable for most genres of rock and pop music.
>
> **POW-r #3:** Even more noise shaping is used. This can extend the dynamic range by 20db in the 2-4khz range. Features psychoacoustically-optimized noise-shaping designed for full-spectrum, wide-stereo field material. This option is recommended for material with a broad dynamic range, such as classical and orchestral music; can also be effective for rock and pop music.
>
> **UV22HR:** This setting uses Apogee's renowned UV22 dithering algorithm, which is said to allow for the best possible sound resolution when bouncing 24-bit recordings to 16-bit files, without adding to the effective noise.

Considerations for Bouncing Audio

When exporting a mix from Logic Pro X, the bounced file will capture all audible information in your mix just as you hear it during playback.

The following principles apply to the mixdown file:

- **The file will include only audible tracks.** What you hear during playback is exactly what will be included in the mix. Any tracks that are muted will not be included. Similarly, if any tracks are soloed, they will be the *only* tracks included.

- **The file will be a rendered version of your project.** Inserts, sends, and external effects will be applied permanently. Listen closely to your entire project prior to exporting the mix to ensure that everything sounds as it should. Pay close attention to levels, being sure to avoid clipping.

- **The file will be created based on the active selection.** If you have an active selection when you export your mix, the mix file will last for the length of the selection only. If you do not have a selection, Logic Pro X will create a bounce from the start of the project to the end of the longest track.

Creating a Bounce of an Audio Mix

When you are ready to create your mixdown in Logic Pro X, you can use the **BOUNCE > PROJECT OR SELECTION** command from the **FILE** menu. You can alternately use the shortcut **COMMAND+B**. This provides a fast and easy way to create a bounce to a stereo file, requiring little to no setup.

Selecting Bounce Options

The Bounce command combines the outputs of all currently audible tracks to create a new audio file on your hard drive. You can select options for the bounced file using the Bounce dialog box.

To bounce all currently audible tracks and select the export options, do the following:

1. Verify that the project plays back as desired. (Check levels, processing, and mute/solo states for tracks).

2. Choose **FILE > BOUNCE > PROJECT OR SELECTION**. The Bounce dialog box will appear.

Figure 9.17 The Bounce dialog box

3. Select one or more of the Destination formats. Once these are chosen, bounce options for that format will appear to the right of the Destination area.

4. Double-check the **START** and **END** fields. These represent the length of your bounce. If you don't want to include the entire project, you can adjust the Start and End fields with the up/down arrows.

5. Under Mode, select **REALTIME** or **OFFLINE**. Offline mode provides a faster-than-realtime bounce; however, it cannot be used with External MIDI sounds. Make sure to select Realtime mode if you're using any External MIDI sounds routed into the Mixer via aux tracks.

6. Two check-boxes are available for additional control over effect tails:

    - **Bounce 2nd Cycle Pass**—This option can be used when bouncing a cycle range. The range will playback for two cycle repetitions, with the bounce starting on the second pass. If you have time-based effects in your project, you will hear the effect bleeding into the start of the bounce.

    - **Include Audio Tail**—This setting extends the end of the bounce to let any time-based effects ring out until they decay. This is recommended when bouncing a project with reverb or delay effects.

7. Select a File Format for your destination file (WAV, AIFF, CAF).

8. Choose the file type for your stereo bounce from the **FORMAT** pop-up menu:

    - **Interleaved**—Creates a single file containing both channels of a stereo mix. Interleaved files are compatible with most online services and music applications, including SoundCloud and iTunes.

    - **Split**—Creates two separate mono files for a stereo mix: one for the left channel and another for the right channel. This file format is required by certain media applications.

9. Choose the desired bit depth for the bounced file(s) from the **BIT DEPTH** pop-up menu:

    - Choose 16 Bit if you plan to burn your bounce to CD without further processing.

    - Choose 24 Bit when you want to create a final mix that will be mastered separately.

10. Choose the desired sample rate for the bounce files from the **SAMPLE RATE** pop-up menu. Higher sampling rates will provide better audio fidelity but will also increase the size of the resulting file(s).

> If you plan to burn your bounced audio directly to CD without further processing, choose 44.1 kHz as the sample rate for the bounce.

11. If desired, enable the **ADD TO PROJECT** checkbox. This option will import the bounced file into your project Audio Browser.

> The Add to Project option is available only if the target sample rate for the bounce file matches the sample rate of your project.

12. Choose a Normalization setting from the **NORMALIZE** pop-up menu.

    - **Off**—No normalization is applied to the bounce.

    - **Overload Protection Only**—Normalization is applied for any levels that exceed 0db, but no normalization takes place for lower levels.

    - **On**—The project's audio is scanned for the highest amplitude peak, then the level is increased so that the peak is at the maximum possible level before clipping.

13. After confirming your settings, click the **OK** button. The bounce process will proceed and a Save dialog box will appear.

    By default, the bounced file will be named after the output channel strip and placed in the **BOUNCES** folder inside the **~/MUSIC/LOGIC** folder for your user account.

14. You can optionally specify a different file name and location for the bounced file using the Save dialog box.

15. When you're ready, click the **BOUNCE** button at the bottom of this dialog box.

Monitoring Bounce Progress

When performing an offline bounce, Logic Pro processes the bounce without audio playback. A window will appear displaying a progress bar during the bounce. (See Figure 9.18.)

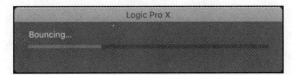

Figure 9.18 The bounce progress window

Locating the Bounced File

After the bounce completes, you can retrieve your mix file from within the **BOUNCES** folder or your selected save location.

To locate your bounced mix, do the following:

1. Using the Application Switcher or the Dock, switch from Logic Pro X to the Finder.

 The Application Switcher lets you toggle through open applications on a Mac by pressing COMMAND+TAB.

2. Navigate to the Bounces folder inside your project folder.

 The default location for project folders is ~/Music/Logic Pro. Your individual project folder locations may be different if you changed the save destination.

Review/Discussion Questions

1. What are some mix attributes that are recalled with a project in Logic Pro X? What are some items that may not be recalled? (See "Recalling a Saved Mix" beginning on page 220.)

2. How many automation modes are available in Logic Pro X? What is the difference between Off mode and Read mode? (See "Available Automation Modes" beginning on page 221.)

3. Which automation mode is useful for writing automation to selected parameters across long sections of your mix? Which mode is useful for touching up automation and making changes that affect only small areas? (See "Writing Real-Time Automation" beginning on page 223.)

4. What is the purpose of displaying an automation curve? What are two ways automation curves can be displayed in Logic? (See "Viewing Automation Curves" beginning on page 224.)

5. What are some ways of editing an automation graph with the Pointer tool? (See "Using the Pointer Tool for Automation" beginning on page 225.)

6. How can the Pencil tool be used to create automation? (See "Using the Pencil Tool for Automation" beginning on page 226.)

7. What is meant by the term *mixing down*? What other term is commonly used to describe this process? (See "Creating a Mixdown" beginning on page 228.)

8. What process can you use to optimize the output levels for your mix? What plug-in can you use in Logic Pro X for this purpose? (See "Adding a Limiter to Optimize Output Levels" beginning on page 228.)

9. What is the purpose of adding dither during a bounce? When would you not want to use dither? (See "Adding Dither to a Mix" beginning on page 230.)

10. What are some things to consider when preparing to create a bounce of your mix? How will the bounce be affected by soloed or muted tracks? (See "Considerations for Bouncing Audio" beginning on page 232.)

11. What sample rate and bit depth should you use when exporting a mix for use on an audio CD? (See "Selecting Bounce Options" beginning on page 232.)

12. Where will your exported file be located after completing the Bounce > Project or Selection command? (See "Locating the Bounced File" beginning on page 235.)

To review additional material from this chapter and prepare for certification, see the Logic Audio Production Basics Study Guide module available through the Elements|ED online learning platform at ElementsED.com.

Exercise 9

Preparing the Final Mix

🎧 Activity

In this exercise, you will perform the steps necessary to create a final mix for your project in Logic Pro X. You'll begin by adding processing to the Stereo Out channel to optimize levels and prepare for bouncing the mix. Then, you'll create a fadeout at the end of the song using real-time automation.

🕓 Duration

This exercise should take approximately 20 minutes to complete.

⊕ Goals/Targets

- Add a limiter to the Stereo Out channel
- Add dither to the Stereo Out channel
- Work with Latch automation
- Save your work for use in Exercise 10

Exercise Media

This exercise uses media files taken from the song, "Lights," provided courtesy of Bay Area band Fotograf.

Written by: Zack Vieira and Eric Kuehnl; Performed by: Fotograf

The media provided for this course may be used for educational purposes only. No rights are granted to use the media for any other personal, commercial, or non-commercial purposes.

Getting Started

To get started, you will open your completed project from Exercise 8. This will serve as the starting point for this exercise.

Open your existing Lights project:

1. Select **FILE > OPEN** or press **COMMAND+O**. The Open dialog box will display.
2. Navigate to the location where you saved your project in Exercise 8.
3. Select your Lights03-xxx project and click **OPEN**. The project will open as it was when last saved.
4. If needed, reopen the Mixer window by pressing the **X** key.
5. Choose **FILE > SAVE AS** and create a copy of the project called Lights04-xxx, keeping the project in the original project folder.

Adding Processing to the Stereo Out Channel

In this section of the exercise, you will add some processing to the Stereo Out channel. First, you'll add a limiter to the channel to optimize the output levels. Then you'll add dither to preserve the best possible sound quality when outputting the mix at 16-bit resolution.

Add a limiter to the Stereo Out track:

1. In the Mixer, click to add an Audio FX insert on the **Stereo Out** track and select **DYNAMICS > LIMITER > STEREO**. The plug-in window will display.
2. Adjust the Limiter settings as follows:
   - Lookahead: 5.0 ms
   - Gain: 3.0 dB
   - Release: 2.0 ms (minimum setting)
   - Output Level: −1.5 dB

Figure 9.19 The Limiter with the suggested settings

3. Press the **SPACEBAR** to begin playback.

4. Watch the Gain Reduction meter (**Reduction**) to see where the limiter is being activated. Allow playback to continue through the first half of the verse.

5. Reset the **Stereo Mix** track to its default level by **OPTION-CLICKING** on the volume fader. You should now see more gain reduction occurring in the plug-in but no clip indication on the track's meters.

6. Continue monitoring playback through the first chorus to observe the results during the loudest sections and verify that the mix is not clipping.

7. Close the plug-in window and stop playback when finished.

Writing Automation

The music for this song has a sudden and unnatural cutoff at the end. You can address the problem by adding a final fadeout at the end of the song. Here, you'll use Latch automation mode to write a fadeout in real time using the volume fader control on the Stereo Out channel.

By default, the Stereo Out channel is not represented in the main window. In order to show its automation curve, you'll need to display it as a track in the main window.

Display the Stereo Out track:

1. Right-click on the Stereo Out channel in the Mixer and select **SHOW OUTPUT TRACK**, or press **COMMAND+SHIFT+M**.

Figure 9.20 The right-click menu on the Stereo Out track in the Mixer

2. If needed, close the mixer by pressing **X** to get a better view of the Tracks area. Note that the Stereo Out track is now displayed.

3. Click the **SHOW AUTOMATION** button in the main window just above the track list, or press the **A** key, to activate the Show Automation function.

4. Click on the Automation Mode selector for the **Stereo Out** track in the main window and choose **LATCH** automation mode.

Figure 9.21 Enabling Latch mode on the Stereo Out track

5. In the Automation Parameter pop-up menu select **MAIN > VOLUME > ABSOLUTE**. (See Figure 9.22.)

Figure 9.22 Selecting volume automation from the automation mode pop-up menu

Automate a fadeout on the Stereo Out track:

1. Press the **SPACEBAR** to begin playback.

2. Click on the horizontal volume slider for the **Stereo Mix** track and slowly move the slider all the way to minus infinity (-∞). You'll have about 10 to 15 seconds to complete the fadeout.

3. Keep an eye on the Counters in the Transport window during your fadeout. Be sure to complete the fadeout before you reach Bar 101. Once you have faded all the way down, you can release the slider and it will stay in position.

4. Stop playback once automation has been written a few bars past the end of the song. The automation curve will show the results of the fadeout you wrote for the track.

Figure 9.23 The volume automation graph on the Stereo Out track showing the final fadeout

5. Press the **SPACEBAR** to listen to the fadeout and verify the automation.

6. If you're not satisfied with the result, try another pass of writing the automation. It frequently takes several attempts to get the timing of a fadeout just right!

7. Once you're happy with the fadeout, set the automation mode on the track back to **READ** mode.

To complete this exercise, you will need to save your work and close the project. You will be reusing this project in Exercise 10, so it's important to save the work you've done.

Finish your work:

1. Press **RETURN** to position the playhead at the beginning of the project.

2. Choose **FILE > SAVE** to save the project.

3. Choose **FILE > CLOSE PROJECT** to close the project.

That completes this exercise.

Chapter 10

Beyond the Basics

...*What to Explore to Become a Power User*...

In this chapter, we take a brief look at various advanced features in Logic Pro X. We begin with a quick overview of using Flex Time and Pitch to manipulate the rhythmic and melodic attributes of audio regions. Then we explore some fundamental MIDI editing techniques that can be used to create and modify MIDI data. Next we discuss how you can simplify project management with submixing. The chapter concludes with a sample of some keyboard shortcuts that you can use to dramatically accelerate your editing workflow.

Learning Targets for This Chapter

- Understand the basics of Flex Time and Pitch
- Explore fundamental MIDI editing techniques
- Gain familiarity with submixing techniques
- Learn essential keyboard shortcuts

Key topics from this chapter are illustrated in the Logic Audio Production Basics Study Guide module available through the Elements|ED online learning platform. Sign up at ElementsED.com.

For those who are new to the world of DAWs, or simply new to Logic Pro X, the information required to produce a finished project can seem a bit overwhelming. Learning the fundamentals should obviously be the first goal, so that you'll have a strong foundation. But it's also helpful to learn some advanced features so you'll be able to work quickly and effectively.

The goal of this chapter is to present some key features in Logic Pro X that you can get started using right away. Many of the topics we touch on here are easily deep enough to warrant an entire chapter in an advanced course. However, here we address just the primary functionality so that you'll be able to put these features to use quickly. Once you've gained a bit of experience with Logic Pro X, you can come back to these topics at a later date and learn them comprehensively.

Flex Time and Pitch

Flex Time is the name of the Logic Pro X feature that allows you to manipulate the timing of notes, beats, and other events in audio regions. A related feature, Flex Pitch, allows you to manipulate the pitch of audio material. Flex operations are not enabled in Logic Pro X by default. You can enable or change the Flex processing on individual tracks in your project at any time.

 Flex Time operations in Logic Pro X are similar to Elastic Audio or Time Warping operations in other DAWs.

With Flex enabled, you have a number of options available to alter the timing and pitch of your regions.

Enabling Flex

Flex Time and Pitch can be accessed in the Tracks area of the Main window and in the Track Editor. You can use one of several methods to activate Flex.

To enable Flex in the Main window, do the following:

1. To show Flex parameters, do one of the following:

   - Choose **EDIT > SHOW FLEX PITCH/TIME** from the Tracks area menu bar.
   - Click the **SHOW/HIDE FLEX** button in the Tracks area menu bar.
   - Press **COMMAND+F**.

Figure 10.1 The Edit menu and Show/Hide Flex Pitch/Time button in the Tracks area menu bar

A Track Flex button and Flex pop-up menu will appear in the track header for each audio track.

2. On the target track, click the **TRACK FLEX** button in the track header to enable Flex for the track. Alternatively, you can select multiple tracks and click one of the **TRACK FLEX** buttons to enable Flex for all selected tracks.

To enable Flex in the Track Editor, do the following:

1. Select the desired track.

2. Click the Editors button (or press the **E** key) to open the Track Editor.

3. In the Track Editor, click the **SHOW/HIDE FLEX** button (or press **COMMAND+F**). A dialog box will display prompting you to turn on Flex for the selected track.

Figure 10.2 The Show/Hide Flex button enabled

4. Select **Turn on Flex** to enable Flex on the track.

Flex Time

Once you've enabled Flex parameters on a track, you can begin to manipulate the timing of the audio on that track. Regions or portions of a region can be expanded or compressed to fit your editing needs. Logic Pro X provides a number of options for editing the timing in audio regions.

The available Flex algorithms include the following:

- **Monophonic**—This algorithm is ideally suited to monophonic material such as vocals or bass guitar when only single notes are played at a time.

- **Slicing**—This algorithm cuts audio into slices at transient markers. This is a good option for drums, percussion, and other material with clear breaks in audio. Slices are played at their original speed, so no expansion or compression occurs.

- **Rhythmic**—This algorithm is the best choice for material that includes rhythm guitars, keyboard parts, and other complex percussive elements.

- **Polyphonic**—This is an all-purpose algorithm that works well on complex material such as chords and mixes. Polyphonic is the most processor intensive of the algorithms.

- **Speed (FX)**—This algorithm links time and pitch changes to create tape speed effects. Stretching audio for slower playback results in lower pitch. Compressing audio for faster playback results in higher pitch.

- **Tempophone (FX)** —This algorithm creates an effect similar to granular synthesis with sound artifacts.

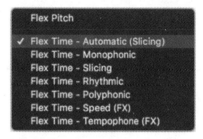

Figure 10.3 The Flex options in Logic Pro X

 When Flex is initially enabled on a track, Flex Time – Automatic is selected by default. In this mode, Logic Pro X analyzes the audio and chooses a suitable algorithm based on the track content.

Flex Time Quantization

The fastest way to tighten the rhythm of an audio performance is to use standard Logic Pro X quantize features. This uses the same settings found in the Region Inspector that are commonly used to alter MIDI performances. The Quantize settings can be used to edit audio data as long as the desired track or regions are selected.

> ### What Is Quantization?
>
> Quantization is the process of aligning musical events to an underlying rhythmic pattern. This process is commonly used to improve the timing of MIDI performances by moving individual MIDI notes closer to the selected grid resolution. This results in more precise timing and a tighter rhythmic feel.
>
> Extreme quantization can make a performance sound "mechanical," which can be desirable for certain genres of music such as EDM. For a more "human" feel, you can apply a subtler amount of quantization and include a degree of strength, randomization, swing, or other characteristics.

To quantize audio using the Quantize dialog box:

1. Verify that Flex Time is enabled on the target track.
2. Select the audio region or regions that you would like to quantize.
3. From the Quantize pop-up menu of the Region Inspector, select a quantization value. Once a value is selected, quantization will be applied to the selection. (See Figure 10.4.)

Figure 10.4 Quantize value set to a quarter note in the Region Inspector

 You can change the quantization settings anytime by selecting another time value. You can also disable quantization by selecting "off."

Manually Edit Flex Markers

If you need to change the timing of your tracks without quantizing to a grid, you have the option of manually editing flex markers. When you show Flex, audio waveforms will show transient markers, and the pointer will turn into a Flex tool. With the pointer, you can turn transient markers into flex markers or add flex markers anywhere within the region. Once placed, these markers can be adjusted by dragging with the pointer. Flex markers can be adjusted or deleted at any time.

Additionally, clicking with the pointer in a region will result in two different behaviors, depending on where the pointer is positioned. Clicking with the pointer in the upper half of a region will place just one flex marker. Clicking with the pointer in the lower half of a region will add a flex marker at that position as well as at the transient markers on either side of the marker.

To add flex markers to change the timing of audio:

1. Verify that the Flex Time is enabled for the target track.

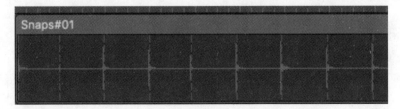

Figure 10.5 A region with Flex enabled showing transient markers

> ⓘ Logic Pro X's Flex Time function adds transient markers automatically at all transient locations and other significant rhythmic events within a region.

2. If no flex marker exists at the location that you'd like to stretch, create one by placing the pointer over the lower half of the region. The pointer will change into a flex tool that shows three flex markers.

Figure 10.6 The Flex tool showing three flex markers when positioned on the lower half of the region

> ⓘ Flex markers anchor the audio at a given point on the timeline. Flex markers are recognizable in Flex view by a blue pentagon at the top and a solid white line.

3. Click on the transient marker. Three flex markers will be added: one at the location, one at the transient before, and one at the transient after.

4. Use the pointer to reposition the flex marker in the center to adjust the timing of the underlying audio event relative to the events on either side.

 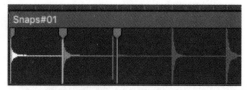

Figure 10.7 A region with flex marker before time stretching (left) and after (right)

> ⓘ When flex markers are adjusted to stretch time, the waveform color will change to reflect the adjustment. When audio is compressed, the waveform becomes white. When audio is expanded, the waveform becomes gray.

Flex Time and the Marquee Tool

Another option for editing time with Flex is to use the Marquee tool. You can reposition a selection rather than stretching the audio between markers. This helps to preserve the original audio while still making time corrections. Just like with the Slicing algorithm of Flex Time, audio that has clear separation between transients will produce the best results.

Similar to adding flex markers as described above, placing the pointer in the upper or lower portions of a region will produce different results. When clicking the upper half of a marquee selection, four flex markers are made: two markers outside the borders of the marquee selection and markers on transients before and after the selection. Clicking the lower half of the marquee selection adds three flex markers: two markers outside the borders of the marquee selection and one at the clicked position within the selection. The former allows you to move the selection like a slice, while the latter stretches the audio within the selection.

To move a marquee selection with Flex:

1. Verify that the Flex Time is enabled for the target track.

2. Make a selection with the Marquee tool.

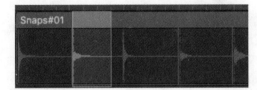

Figure 10.8 A marquee selection on a Flex-enabled region

3. Position the pointer over the upper half of the selection and click. (The pointer will turn into a hand.) Four flex markers will be created: two on the borders of the marquee selection and one marker on each transient to the left and right of the selection.

4. Grab the marquee selection with the hand and drag to the desired location.

Figure 10.9 A marquee selection with flex markers before editing (left) and after (right)

Flex Pitch

Flex Pitch allows you to alter the melodic material of an audio file. Instead of displaying transient markers, audio content will have an overlay of MIDI notes. When Flex Pitch is enabled, you can adjust the individual notes on a piano roll, much like you would for a software instrument track. Notes can be moved vertically or horizontally, resized, split, and merged.

 Flex Pitch works only for monophonic, or melodic, content. It cannot process individual notes in a chord or other complex sound.

Each displayed note also has six parameters that can be adjusted, independent of the other notes. When you select a note or move the pointer over it, the six parameters, or "hotspots," will appear. From left to right, hotspots along the top are pitch drift at the note start, fine pitch, and pitch drift at the note end; the hotspots along the bottom are gain, vibrato, and formant shift (see Figure 10.10).

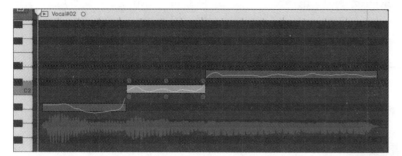

Figure 10.10 Flex Pitch "hotspots" on a selected note in the Audio Track Editor

The fine pitch parameter allows you to fine tune notes. This is useful for when the notes of a performance are correct but are out of tune. Fine pitch makes subtle adjustments so that a performance will better harmonize with other parts. You can tell if a note is in tune visually by how solid the note is. When a note is right on the MIDI note position, it's in perfect pitch. Notes that sit above or below are sharp or flat, respectively (see Figure 10.11).

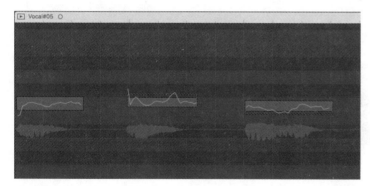

Figure 10.11 Flex Pitch shows whether notes are in perfect pitch (left), sharp (middle), or flat (right)

 When editing notes in the Track Editor, you can easily access additional features like the pitch correction slider and scale quantize menu.

To change pitches of a melody with Flex Pitch:

1. Verify that the Flex Pitch is enabled for the target track.
2. Select or position the pointer over the note you want to change.
3. Drag notes vertically to the desired pitch. (See Figure 10.12.)

Figure 10.12 Lowering the third note of the example in Figure 10.10 with Flex Pitch

4. Adjust notes that aren't in perfect pitch using the fine pitch hotspot: select the note and drag the hotspot up or down until it resets to 0, or other value you choose.

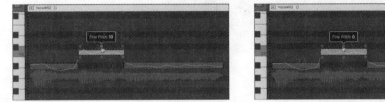

Figure 10.13 Using Fine Pitch parameter to adjust the sharp note (left) and make it perfect pitch (right)

MIDI Editing Techniques

MIDI editing is another topic worthy of its own chapter (or several chapters). But as with Flex, a little knowledge is all you need to get started. Understanding the fundamentals of MIDI editing is absolutely essential in the modern studio. Whether your goal is to add a few MIDI instruments to a song or to compose a virtual orchestral masterpiece, you'll need to know how to edit MIDI data quickly and accurately.

Viewing MIDI Regions and Data

When working with MIDI-capable tracks, the two areas you'll utilize are the Tracks area and Editors pane or window. The Tracks area of the Main window allows viewing and editing of MIDI regions only, making it useful for arranging a project. The Piano Roll Editor, on the other hand, allows you to view and edit regions but also provides full access to MIDI information contained within regions, making it a more versatile option. The Piano Roll can be opened in the Editor pane below the Tracks area or as a separate window. Both viewing options give you the same features.

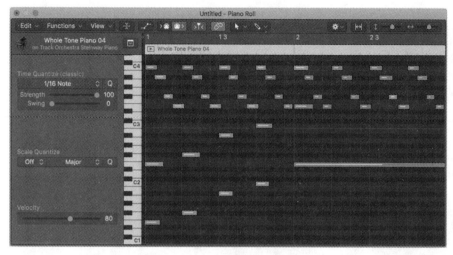

Figure 10.14 Piano Roll window

MIDI Regions in the Tracks Area

On MIDI-capable tracks, the Tracks area of the Main window is commonly used to edit and arrange MIDI regions. In this area, you can perform the same basic editing techniques that you use on audio regions:

- Cut, Copy, Paste, and Clear selections or regions
- Duplicate and Repeat selections or regions
- Trim regions

Figure 10.15 A Software Instrument MIDI region in the Tracks area

> ### Repeat Commands
>
> The two commands, Repeat and Repeat Multiple, provide ways to quickly create multiple instances of a region or selection. Repeat is the best option when you only need to make one or two sequential copies. If you need to make multiple copies, it can be faster to use the Repeat Multiple command instead.
>
> To use the Repeat command, select a region or a portion of a region. Then choose **EDIT > REPEAT** or press **COMMAND+R**. To use the Repeat Multiple command instead, choose **EDIT > REPEAT MULTIPLE** and type in the number of copies to make.

MIDI Regions in the Piano Roll Editor

The Piano Roll Editor is where you'll do most of your work on MIDI regions. You can perform a wide range of tasks using this window, including the following:

- Adding and deleting notes
- Selecting notes
- Transposing notes (changing a note's pitch)
- Moving notes (changing a note's timing)
- Trimming the note start and end (changing a note's duration)
- Quantizing notes

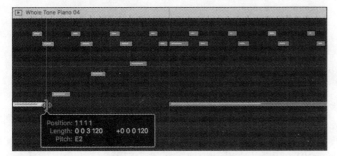

Figure 10.16 Trimming a MIDI note in the Piano Roll Editor

Automation/MIDI Area

The Automation/MIDI area in the Piano Roll Editor can be used to edit the velocity and other controller data associated with individual MIDI notes. A note's velocity is a measurement of how hard the note has been struck. This is one of the most common pieces of data you will edit along with MIDI notes. The velocity characteristic can substantially change the sound of a virtual instrument.

Automation and MIDI data is show in two ways: as an automation curve and as nodes with horizontal lines representing note lengths. Values can be adjusted in a number of ways, including:

- Adjusting the velocity of a note with the pointer
- Scaling velocity values of multiple notes by dragging up or down with the Pencil tool
- Drawing gradual changes in velocity (crescendo/decrescendo) with the pointer

Figure 10.17 Adjusting the velocity value of a note in the MIDI/Automation area of the Piano Roll Editor

Editing MIDI Data

Logic Pro X offers a powerful MIDI editing toolset that can be used to create and edit MIDI performances, regardless of whether or not you own a MIDI controller such as a keyboard.

Let's take a closer look at some options for creating and working with MIDI data in Logic Pro X.

Creating MIDI Notes

Manually entering MIDI note data may seem like a slow method of creating a performance, compared to playing the performance on a MIDI controller. However, many music producers prefer to manually enter note data for certain types of tracks (such as drums, percussion, and bass). With a little practice, you can create and edit MIDI note data very quickly and accurately, using only the Pencil tool and the Pointer tool.

To manually create a MIDI note:

1. Select the Software Instrument track you wish to use.

2. If there is no MIDI region to edit, create one by **CONTROL-CLICKING** an empty area in the Tracks area or Piano Roll Editor. An empty one-bar region will be created.

3. In the Piano Roll Editor, hold **COMMAND** to change the pointer to the Pencil tool.

4. Click and hold with the Pencil at the location (in both pitch and time) where you would like to enter a note. You can adjust the note's duration by dragging forward/backward; you can adjust the note's pitch by dragging up or down.

> ⓘ If no MIDI region exists in the Piano Roll Editor, clicking anywhere in the grid with the Pencil tool will automatically create a new region along with a note at the clicked location.

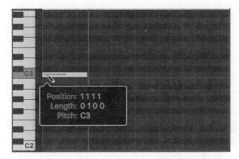

Figure 10.18 Manually entering a MIDI note with the Pencil tool

5. Release the mouse to complete the action and place the note at the selected location.

 Newly created notes match the length, velocity, and channel of the previously edited or selected note. When opening a new project, the default values are: length equals 240 ticks (sixteenth note), velocity equals 80, and MIDI channel is 1.

Selecting MIDI Notes

Another essential part of working with MIDI data involves selecting notes. Understanding how to quickly select single notes and groups of notes is critical to the MIDI editing workflow. You can use the following techniques to select notes on a MIDI or Instrument track in the Piano Roll Editor.

To select an individual note, do one of the following:

- Using the pointer or Pencil tool, click anywhere on a note.
- Using the pointer, click in an empty space on the grid and drag a rectangle over the note.

To select groups of notes, do one of the following:

- Using the pointer or Pencil tool, **SHIFT-CLICK** on each of the desired notes.
- Using the pointer, click in an empty space on the grid and draw a rectangle around the desired notes.

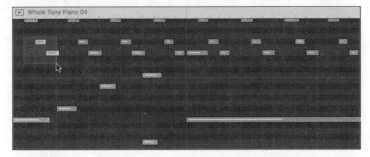

Figure 10.19 Selecting a group of MIDI notes with the pointer

Editing MIDI Notes

With basic note creation and selection techniques in hand, a whole range of MIDI editing functions are available to you. Let's look at some essential MIDI editing techniques that you can use.

Deleting MIDI Notes

To delete an individual note, do one of the following:

- Using the pointer, select a note and press **DELETE**.
- Using the Pencil tool, **DOUBLE-CLICK** on a note.

To delete a group of notes:

- Select the notes and press **DELETE**.
- Select the notes, then using the Pencil tool, **DOUBLE-CLICK** on a selected note.

Moving MIDI Notes

To move a note to a new location, do one of the following:

- Using the pointer or Pencil tool, click on the note and drag left or right.
- Select a note and press **SEMICOLON** (;) to move the selected note to the playhead position. (This is particularly useful when moving to a location off screen.)
- Select a note, then press **OPTION+LEFT ARROW** to nudge the note to the left or **OPTION+RIGHT ARROW** to nudge to the right.

> Nudging is an editing technique that allows you to move a selected event or region forward or backward in time by small increments using key commands.

> To change the nudge value, open the Toolbar by clicking the Toolbar button in the Control bar or by pressing Control+Option+Command+T. From the Nudge Value pop-up menu, select the desired increment.

To move a group of notes to a new location, do the following:

1. Select the target notes.
2. Use any of the methods listed above for moving a single note.

When working in Grid mode, hold Control to temporarily suspend the grid. For finer control, hold Control+Shift.

Transposing MIDI Notes

To change a note's pitch, do one of the following:

- Using the pointer or Pencil tool, click on the note and drag up or down.
- Select a note and press **OPTION+UP ARROW** to transpose up a semitone or **OPTION+DOWN ARROW** to transpose down a semitone.

 To transpose an octave up or down, hold SHIFT+OPTION+UP ARROW or SHIFT+OPTION+DOWN ARROW, respectively..

To change the pitch of a group of notes, do the following:

1. Select the target notes.
2. Use any of the methods listed above for transposing a single note.

 When moving or transposing notes with the pointer or Pencil tool, you can hold Option to copy the notes. This is a quick way to create harmonies or chords for melodic instruments.

Trimming MIDI Notes

To trim a note, do the following:

1. Using the pointer or the Pencil tool, position the cursor near the beginning or end of the note.
2. Click and drag left or right to trim the note start or end point.

To trim a group of notes, do the following:

1. Select the target notes.
2. Using the pointer or the Pencil tool, position the cursor near the beginning or end of any one of the selected notes.
3. Click and drag left or right to trim the start or end point of all of the selected notes.

When trimming multiple notes of different lengths, hold Shift+Option to make them the same length as the note being trimmed.

Working with Velocity

Most modern MIDI controllers are "velocity sensitive," meaning that they automatically detect how hard a key or pad is struck. The velocity data is transmitted and recorded as a characteristic of the MIDI note. However, when notes are manually entered (with the Pencil tool, for example), they will all share a default velocity value. This can lead to very stiff, robotic–sounding performances. Modifying and manipulating velocity data is essential to creating natural-sounding MIDI parts.

We've already covered how to edit velocity in the Automation/MIDI area of the Piano Roll Editor. Two more options are available that are quick and easy alternatives. The first is to use the Velocity slider to the left of the Piano Roll and below the quantize settings. The second is to use the Velocity tool.

To edit velocity using the Velocity slider:

1. With the pointer or Pencil tool, select the desired note.

2. Move the Velocity slider to the preferred setting.

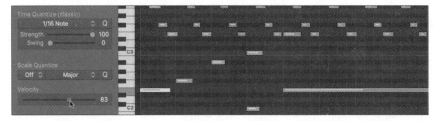

Figure 10.20 Adjusting velocity of a note with the Velocity slider

To edit velocity using the Velocity tool:

1. Select the Velocity tool from the Left-click Tool menu.

2. Click and hold on a note and drag up or down to the preferred setting.

To edit MIDI velocity data for a group of notes:

1. Select the group of notes that you wish to affect.

2. Use either technique described above.

In addition to the Velocity slider and Automation/MIDI data area, Logic Pro X shows velocity visually on the notes in two ways. First, horizontal lines within the notes represent the note's relative velocity. Second, the note color changes with velocity, ranging from violet (low velocity) to red (high velocity).

Software Instrument Region Inspector

During the music production process, you'll frequently want to make adjustments to MIDI data. One available option is to manually transpose, set quantization, and otherwise fine-tune a MIDI performance with various tools.

However, when you're in the middle of the creative flow and you want to perform a simple quantize operation, it's not always convenient to stop and look at a complex dialog box or specific tool. In these instances, you can take advantage of the Region Inspector settings in Logic Pro X.

The Region Inspector controls can be used to quickly apply a number of settings, including the following modifications to the regions on a MIDI-capable track:

- **Quantization**: Select a quantize grid value and swing amount.
- **Q-Strength**: Adjust how far a note gets shifted toward the grid.
- **Delay**: Adjust the timing of notes, shifting them earlier or later.
- **Velocity**: Scale the velocity of notes up or down or change the range of velocities by percentage.
- **Transpose**: Transpose notes up or down by a specified number of octaves and/or semitones.

A powerful advantage of using the Region Inspector is that you can quickly toggle a number of the properties on and off at any time. And you can modify the settings at any time without having to undo a previous change.

To view the full list of Region Inspector settings:

1. With a software instrument track selected, click the **INSPECTOR** button or press **I** to open the Inspector.
2. Click to expand the **MORE** disclosure arrow. Once expanded, the full list of parameter controls for the region will appear. You can use these controls to quickly adjust the settings.

Figure 10.21 The Region Inspector settings on a Software Instrument track

Quantization in the Piano Roll Editor

One of the most commonly used tools is quantization. Earlier in this chapter, we looked at the quantization features in the Region Inspector when editing and correcting audio with Flex Time algorithms. MIDI-capable tracks share many of the same quantization settings in the Region Inspector, but also have the most common options conveniently available within the Piano Roll Editor: Time Quantize and Scale Quantize.

An added benefit of using the Piano Roll Editor for quantization is that you can select specific notes rather than operating on all content within an entire region.

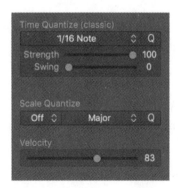

Figure 10.22 The quantization settings in the Piano Roll Editor

The Time Quantize settings can be used to quickly apply the following:

- **Time Quantize pop-up menu:** Set a quantize value to apply correction. The default is 1/16 Note.
- **Time Quantize (Q):** Apply correction to selected notes based on the set Time Quantize value.

- **Strength:** Set the degree of quantization applied to a selection (default is 100).
- **Swing:** Set the degree of swing (default is 0).

The Scale Quantize settings can be used to quickly limit notes to a particular scale. This function uses the following parameters:

- **Scale Quantize pop-up menu:** Enable and disable scale quantization from the left menu (default is OFF). Select the scale from the right menu.
- **Scale Quantize (Q):** Apply correction to selected notes based on the set Scale Quantize settings.

 Quantization applies correction to selected events without permanently altering the content. You can always return the content to its original state at any time.

Submixing with Track Stacks

Submixing is a technique whereby multiple tracks are summed to a common destination (typically an aux channel) to simplify certain mixing tasks. While submixing is most useful in projects with a large number of tracks, there are advantages to using submixes in smaller projects as well.

Simplifying a Mix

Submixing is frequently used to help simplify the mixing process. By submixing a group of related tracks, you can more easily apply effects across the whole group. You can also automate the summed levels of the group using a single fader.

Submixing is an essential technique for managing drum recordings, which typically span across a number of tracks. By routing the drums to a common Aux Input, you can apply compression or limiting to the group (for example) using a single plug-in. Then, you can quickly set the level for the entire drum kit using a single fader, while maintaining the relative levels of the individual tracks, as set by their own faders.

But submixing isn't just for drums. It can be a great way to manage any related group of tracks such as background vocals, guitars, and keyboards.

Using Track Stacks to Submix

Logic Pro X offers a quick and convenient way to submix, using a feature called Track Stacks. This feature groups numerous tracks together into a collapsible display within the Tracks area and Mixer window. The required submix signal routing is handled automatically.

Submix workflows utilize a type of Track Stack called a *summing stack*. Summing stacks combine the outputs from multiple tracks and route the summed output to a bus functioning as an audio subgroup. The subgroup is represented by a top-level track known as the *main track*. This track is assigned to an aux channel strip in the mixer.

 Logic Pro X provides two types of Track Stacks: folder stacks and summing stacks. This discussion focuses on summing stacks.

By selecting the main track for a summing stack, you can mute, solo, and adjust volume and send levels for the stack. You can also add plug-ins to the main track, affecting the sound of all the subtracks in the summing stack.

Summing stacks are ideal, not only for adding processing to a group and controlling its overall level, but for simplifying the organization within your project.

To create a Track Stack for submixing, follow these steps:

1. Select the tracks you wish to submix.

2. Choose **Track > Create Track Stack** or press **Command+Shift+D** to open the Track Stack dialog box.

 Use the key command Command+Shift+G to create a summing stack directly, bypassing the dialog box.

3. In the dialog box, select the **Summing Stack** radio button and click **Create**. A Track Stack will be created, and the selected tracks will be moved under the stack's main track.

4. Optionally, rename the main track with a descriptive name for your submix.

Figure 10.23 A drum submix created using a Summing Stack

Once you've created a summing stack, you can collapse or expand the Track Stack at any time by clicking on the disclosure arrow to the left of the track icon on the main track.

You can easily insert plug-ins on the main track to have them apply to the submix. You can also adjust the volume and panning on the main track to modify the submix level and position in the larger mix.

Creating Stems

Another important use for submixes is to aid in the creation of *stems*. Stems are mixed-down versions of your submixes that can be useful when mixing or remixing a project in a different DAW. Stems can also be used to supplement a live performance.

Once you've routed all of your tracks to summing stacks, it is very easy to create stems in Logic Pro X.

To create stems from summing stacks, do the following:

1. Make a selection with Cycle that spans the section of the song that you would like to bounce as stems. This might be the entire song, or it might be one section of the song like a verse or chorus. If no selection is made, the bounce will extend from the beginning until the Project End marker.
2. Locate the first summing stack that you would like to bounce as a stem.
3. Select the main track for the stack.
4. Press **S** or click the **SOLO** button on the main track. This will solo all of the tracks in the stack.
5. Play the session for a bit to verify that no other tracks are currently audible.

> ⓘ If the tracks in the summing stack include sends to a reverb or delay, make sure that the reverb or delay aux channel strip is also soloed.

6. If you are using any plug-ins (such as a limiter) on the Stereo Out channel strip, consider bypassing or removing them.

> ⓘ Generally when using submix tracks in another DAW, you will want to apply master bus processing there for maximum control. However, you can also create stems with processing in place on the Stereo Out channel strip in Logic Pro X to help preserve the sound of your mix.

7. Choose **FILE > BOUNCE > PROJECT OR SECTION** or press **COMMAND+B**. The Bounce dialog box will appear.
8. Choose the desired settings from the Bounce dialog, and then click **OK**. The submix will be exported as an audio file.

> ⓘ Due to the nature of effects like reverb, the effect's decay tails may extend beyond the end point of the selection. This can result in an abrupt sound cutoff at the end of the bounced audio file. As a safeguard, check the box to Include Audio Tail to ensure the audio file continues until volume reaches zero.

9. Repeat the process for each of the summing stacks in your project.

When you are finished, you will have an audio stem file for each of the submixes. These files can then be use in another Logic Pro X project, or in an entirely different DAW. Using these stems, you can recreate the original mix with no more than a few small tweaks.

 If you have no signal processing on aux channel strips or the Stereo Out channel, or if you plan to reapply processing in another environment, you can use the Export function to export one or more selected main tracks as audio files. This can save time for larger projects. The resulting audio files will not contain signals from Sends, aux channels, or the Stereo Out channel.

Keyboard Shortcuts

In closing, we'll cover some keyboard shortcuts that will dramatically accelerate your work in Logic Pro X. Learning even a handful of essential shortcuts can elevate your status from prospect to pro user in no time.

 Logic Pro X allows users to edit key commands and create shortcuts for hundreds of commands that are unassigned by default. You can save your custom commands for use on another system or import custom sets. You can even import settings that match other DAWs like Pro Tools.

 To open the Key Command window, choose Logic Pro X > Key Commands > Edit or press Option+K.

Working Areas in Main Window

Key	Function
Y	Show/Hide the Library
I	Show/Hide the Inspector
B	Show/Hide the Smart Controls
X	Show/Hide the Mixer
E	Show/Hide the Editor
O	Show/Hide the Loops Browser
F	Show/Hide the Browsers

Other Windows

Key	Function
COMMAND+N	Open New Project from Template
OPTION+COMMAND+W	Close Project
COMMAND+2	Open Mixer Window
COMMAND+4	Open Piano Roll Window
COMMAND+K	Show/Hide Musical Typing

Zoom Functions

Key	Function
Z	Toggle Zoom Fit Selection to Screen
COMMAND+LEFT ARROW	Zoom Horizontal Out
COMMAND+RIGHT ARROW	Zoom Horizontal In
COMMAND+UP ARROW	Zoom Vertical Out
COMMAND+DOWN ARROW	Zoom Vertical In

Navigation and Playback

Key	Function
SPACEBAR	Play/Stop
SHIFT+SPACEBAR	Play from Selection
RETURN	Return to Beginning
COMMA (,)	Rewind One Bar
PERIOD (.)	Skip Forward One Bar
C	Toggle Cycle Mode On/Off
U	Set Rounded Locators by Selection and Enable Cycle

Tracks

Key	Function
OPTION+COMMAND+N	New Tracks
COMMAND+D	New Track With Duplicate Settings
S	Solo the Track
M	Mute the Track
A	Show/Hide Automation
COMMAND+F	Show/Hide Flex
G	Show Global Tracks
COMMAND+SHIFT+D	Create Track Stack

Edit Commands

Key	Function
COMMAND+Z	Undo
COMMAND+X	Cut
COMMAND+C	Copy
COMMAND+V	Paste
COMMAND+A	Select All
COMMAND+T	Split at Playhead
COMMAND+J	Join Regions
COMMAND+R	Repeat
SEMICOLON (;)	Move to Playhead

Bounce and Export

Key	Function
COMMAND+B	Bounce Project or Selection
CONTROL+B	Bounce in Place
COMMAND+E	Export Tracks as Audio Files
COMMAND+F	Show/Hide Flex

Review/Discussion Questions

1. What windows or areas provide access to Flex functions in Logic Pro X? (See "Enabling Flex" beginning on page 244.)

2. Which Flex Time algorithm is best suited to monophonic material such as vocals or bass guitar? (See "Flex Time" beginning on page 245.)

3. Which algorithm is the best for drums and percussion? (See "Flex Time" beginning on page 245.)

4. Is it possible to use Flex Time to tighten the rhythm of an audio performance? What process would you use? (See "Flex Time Quantization" beginning on page 246.)

5. How can you quickly create three successive Flex markers on a Flex-enabled track, in order to reposition a spot in the center? (See "Manually Edit Flex Markers" beginning on page 247.)

6. What two areas are used to edit MIDI data? What editing operations can you perform in each? (See "Viewing MIDI Regions and Data" beginning on page 251.)

7. What two tools are commonly used for creating and editing MIDI data? (See "Creating MIDI Notes" beginning on page 254.)

8. How can new MIDI notes be manually created on the Piano Roll? (See "Creating MIDI Notes" beginning on page 254.)

9. What are some common techniques for editing MIDI notes? (See "Editing MIDI Notes" beginning on page 256.)

10. What is MIDI velocity? What are some ways that velocity data can be viewed and edited in Logic Pro X? (See "Working with Velocity" beginning on page 258.)

11. What are some Region Inspector controls that can be used to quickly apply changes to the regions on a MIDI-capable track? (See "Software Instrument Region Inspector" beginning on page 259.)

12. How is submixing useful in a project? What process can you use to quickly create a submix in Logic Pro X? (See "Submixing with Track Stacks" beginning on page 261.)

13. What type of Track Stack is used for a submixing workflow? (See "Using Track Stacks to Submix" beginning on page 261.)

14. Why is it helpful to create stems when exporting your Logic Pro X project to be imported into a different DAW? (See "Creating Stems" beginning on page 263.)

 To review additional material from this chapter and prepare for certification, see the Logic Audio Production Basics Study Guide module available through the Elements|ED online learning platform at ElementsED.com.

Exercise 10

Finalizing a Project

🎧 Activity

In this exercise, you will perform the final steps necessary to finish a project in Logic Pro X. You'll begin by creating a drums submix. Then, you'll duplicate and transpose the **Strings** track to achieve a bigger string sound. Finally, you'll bounce your mix to a stereo audio file.

🕓 Duration

This exercise should take approximately 20 minutes to complete.

◈ Goals/Targets

- Submix multiple tracks to a summing stack
- Duplicate a Software Instrument track to create a variation on an existing MIDI performance
- Use Region Inspector controls to transpose MIDI data to a different octave
- Bounce the final mix as a stereo file

Exercise Media

This exercise uses media files taken from the song, "Lights," provided courtesy of Bay Area band Fotograf.

Written by: Zack Vieira and Eric Kuehnl; Performed by: Fotograf

The media provided for this course may be used for educational purposes only. No rights are granted to use the media for any other personal, commercial, or non-commercial purposes.

Getting Started

To get started, you will open your completed project from Exercise 9. This will serve as the starting point for this exercise.

Open your existing Lights project:

1. Select **FILE > OPEN** or press **COMMAND+O**. The Open dialog box will display.

2. Navigate to the location where you saved your project in Exercise 9.

3. Select your Lights04-xxx project and click **OPEN**. The project will open as it was when last saved.

4. Choose **FILE > SAVE AS** and create a copy of the project called Lights05-xxx keeping the project in the original project folder.

Creating a Drum Submix

You'll begin by creating a submix of all the drum tracks by combining them to a single summing stack. You will then be able to use the main track for the summing stack to set the volume for the drums as a whole, without having to adjust the faders on all of the individual tracks.

Create a summing stack for a drum submix:

1. Select the four drum tracks (01 Kick, 02 Snare, 03 Hi Hat, and 04 Claps).

2. Choose **TRACK > CREATE TRACK STACK** or press **SHIFT+COMMAND+D** to open the Track Stack dialog box.

3. Select the **Summing Stack** radio button and click **CREATE**. A main track will be created for the summing stack, and the selected tracks will be moved under the main track.

4. Rename the main track Drum Submix.

Figure 10.24 The renamed main track for the summing stack containing the four drum tracks

You can now take a moment to adjust the level of the drums relative to other tracks in the mix using the volume fader on the **Drum Submix** main track. Try pushing the **Drum Submix** up a few dB to see how it changes the mix.

Adding a Second Strings Track

The **Strings** track may sound a little thin for the mix at this point. In this section, you will help the strings sound more like a real orchestra by adding a second track and transposing that track down an octave.

Duplicate the Strings track and its contents:

1. Click on the **Strings** track and verify that it is the only track selected.
2. While holding **OPTION**, click and drag the **Strings** track to an empty space to duplicate the track and its contents. You will now have a second copy of the **Strings** track and its MIDI data.
3. Rename the new track **Strings Low** and rename the original track **Strings Hi**.
4. Pan the **Strings Low** track to the right a bit. This will be a nice complement to the panning you did in Exercise 7 on the **Strings Hi** track.

Figure 10.25 The duplicated Strings track and its adjusted panning settings

In the next series of step, you'll transpose the duplicated notes to double the strings an octave lower.

Transpose the Strings Low track using the Region Inspector:

1. If the Inspector is not already open, click the Inspector button or press **I**.
2. Select the **Strings Low** region to show its settings in the Region Inspector.
3. From the **TRANSPOSE** pop-up menu, select **-12**. This will transpose the MIDI notes in the **Strings Low** region down by one octave. The region will now display "(-12)" after its name.

Figure 10.26 The Strings Low track with the suggested Region Inspector transposition

Bouncing an Audio Mix

In this section of the exercise, you will bounce the final audio mix of your project. Bouncing a mix is a necessary step whenever you want to burn a song to CD, post it to a music-sharing site like SoundCloud, or sell it on services like iTunes or Bandcamp.

For this project, you'll create a 16-Bit, 44.1 kHz WAV file. Selecting these parameters will ensure that the resulting file is compatible with all of the common distribution options.

Bounce your final mix:

1. In the Tracks area of the Main window, press **COMMAND+A** to select all regions.

2. Select **NAVIGATE > SET ROUNDED LOCATORS BY SELECTION AND ENABLE CYCLE** or press **U** to set Cycle locators at the start and end of the song.

> ⓘ If you don't make a selection before exporting your mix, Logic Pro X will make certain assumptions about where your project ends, which may not be accurate. It's generally a good idea to set the range yourself to control the exact length.

3. Select **FILE > BOUNCE > PROJECT OR SELECTION** or press **COMMAND+B**. The Bounce dialog box will display.

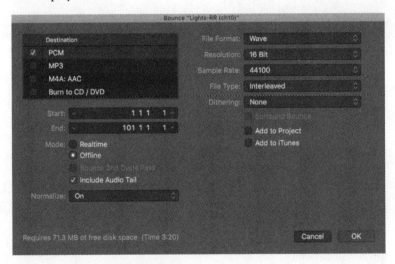

Figure 10.27 The Bounce dialog box

4. In the Bounce dialog box, select the following settings:

    - PCM: Checked
    - Include Audio Tail: Checked

- **File Format:** Wave
- **Resolution:** 16 Bit
- **Sample Rate:** 44100
- **File Type:** Interleaved

5. Select **OK** to confirm the settings.
6. On the next dialog box, specify the filename for your mix.
7. If desired, select a different location to save the file.

> If you do not specify a destination, your mix will be bounced and saved to the Bounces folder located in the project folder.

8. To listen to the mix as the project is being exported, check the **Realtime** radio button. This will create a real-time export. Otherwise, check the **Offline** radio button for a faster-than-real-time export.
9. When ready, click the **BOUNCE** button to begin bouncing your mix. Depending on the option you selected, you will see a countdown or a progress bar as the export completes.

Finishing Up

Congratulations! You've completed all of the required audio production work for a Logic Pro X project, including:

- Importing audio and MIDI files
- Editing clips and assigning virtual instruments
- Mixing and balancing tracks
- Adding EQ, dynamics, and effects processing
- Creating automation
- Submixing tracks
- Applying MIDI Region Inspector parameters
- Bouncing the final mix as a stereo file

To wrap up, you'll need to locate the file you bounced and verify the results.

Locate your bounced stereo mix file:

1. Switch from Logic Pro X to the Mac Finder.

2. If you kept the default export location in the Bounce dialog box, do the following:

   - Locate and open your project folder.

   - Open the **Bounces** folder within the project folder. Your mix file should be within this folder.

   Otherwise, navigate to the folder location you selected for your bounce.

3. Select your mix file and do one of the following to listen to the results of your bounced mix:

   - Press the **SPACEBAR** to begin playback using the QuickLook feature.

   - Open the audio with your preferred media player.

4. Listen through the entire mix to verify that the results are as you expected/intended.

 Be sure to listen to the mixed files that you create with Logic Pro X before sharing them with others. If an unintended condition affects your bounce, you may not realize it without a careful listen, especially if you use the Offline option.

5. If you hear any problem with the file (such as missing parts, wrong file duration, etc.), return to Logic Pro X to correct the issue. Look for soloed or muted tracks, a selection that is too long or too short, and so forth. Once corrected, play back the project to verify that it sounds correct and repeat the bounce process.

6. When happy with the results, return to Logic and save your work.

7. Choose **FILE > CLOSE PROJECT** to close the project.

That completes this exercise.

Index

A

Adaptive Limiter, 202
Advanced Tools, 46
All Files Browser, 146
Analog vs Digital Audio, 66
Apple Loops, 148
Audio Basics, 58
    Amplitude, 59
    Frequency, 58
Audio Interfaces, 68
    Benefits, 11
    Budget, 70
    High-Resolution Audio, 69
    I/O Connectivity, 69
    Sound Quality, 69
    Working Without, 70
Audio Mix
    Bouncing, 232
Automation, 220
    Automation Modes, 221
    Editing Automation Points, 225
    Editing with Pencil, 226
    Editing with the Pointer Tool, 225
    Latch Mode, 221
    Modifying, 224
    Read Mode, 221
    Relative Mode, 222
    Selecting a Mode, 222, 223
    Touch Mode, 221
    Trim Mode, 221
    Viewing Automation Curves, 224
    Viewing Subtracks, 225
    Write Mode, 221
    Writing in Automation in Real-Time, 223

B

Bounce Options, 232
Bouncing
    Adding Dither, 230
    Audio Mix, 232
    Keyboard Shortcuts, 267
    Locating Bounced Files, 235
Bouncing Audio
    Considerations, 232
Bouncing Tracks, 232
Buchla, Don, 78

C

Catch Mode, 124
Catch Settings
    Catch When Starting Playback, 125
Clipping, 169
Color Palette, 175
Compressor, 200, 202
    Strategies, 202
    Using Compression, 203
Control Bar
    LCD, 110
CPU, 6
    Hyper-Threading, 6
    Processing Requirements, 7
    Processing Speed, 6
    Processor Cores, 6
Creating a CD Mix, 230
Creating a Project, 102
Crossfade
    Creating a Crossfade, 157

D

DAW
    Ableton Live, 35
    Apple GarageBand, 33
    Apple Logic Pro X, 32
    Avid Pro Tools, 34
    Avid Pro Tools|First, 34
    Cakewalk Sonar, 37
    Cockos Reaper, 37
    Common DAWs, 32
    Functions, 32
    MOTU Digital Performer, 37
    Presonus Studio One, 37
    Propellerhead Reason, 37
    Steinberg Cubase, 36
    Steinberg Nuendo, 36
De-Esser, 200
    Strategies, 203
    Using, 203
DeEsser 2, 202
Delay, 204
    Applications, 207
    Controls, 206
    Definition, 206
    Usage, 207
Digital Audio
    Analog-to-Digital Conversion, 67
    Binary Data, 66
    Bit Depth, 68
    Dynamic Range, 68
    Sample Rate, 67
Dither, 230
    Noise Shaping, 231
D-Verb
    Controls, 205
Dynamics, 171, 199
    Attack/Release, 202
    Basic Parameters, 201
    Compressors, 200
    De-Essers, 200
    Expanders, 201
    Gates, 201
    Knee, 201
    Limiters, 200
    Ratio, 201
    Threshold, 201
    Types, 200

E

Edit Commands
    Keyboard Shortcuts, 266
    Repeat, 252
    Repeat Multiple, 252
Edit Window
    Rulers, 114
Editing
    Fading Regions, 155
    MIDI Regions, 154
    Non-Destructive, 154
    Splitting a Region, 155
    Trimming, 154
Editing Regions, 149
Editing Techniques
    Advanced, 154
    Basic, 152
Enveloper, 202
EQ, 171, 196
    Basic Parameters, 197
    Filters, 197
    Frequency, 197
    Gain, 197
    Notch, 197
    Parametric, 197
    Q, 197
    Shelf, 197
    Strategies, 198
    Types, 196
    Using, 198
Eraser tool, 120
Expander, 201, 202

Export Options
    Bit Depth, 234
    Format, 233
    Import Exported File, 234
    Sample Rate, 234
Exporting
    Audio Mix, 232
    Bouncing, 232
    Keyboard Shortcuts, 267

F

Fades
    Creating, 156
File Hierarchy, 19
File Management, 12
    Creating Folders, 16
    Locating Files, 15
    Mac OS, 12
    Moving Files, 15
    Opening Files, 15
    Right-Click Functions, 16
    Using File Explorer, 14
    Using the Finder, 12
    Windows, 14
Flex Markers, 247
    Adding, 247
Flex Pitch, 244, 249
    Enabling, 244
Flex Time, 244, 245
    Enabling, 244
    Marquee Tool, 248
    Monophonic, 245
    Moving a Selection, 249
    Polyphonic, 245
    Quantizing, 246
    Rhythmic, 245
    Slicing, 245
    Speed (FX), 245
    Tempophone (FX), 246
Flex tool, 247

G

Gain-Based Processing, 176
Gate, 201
Grid Value
    Setting the Grid Value, 150

H–I

Horizontal Scrolling, 123
Import Command, 147
Importing
    Apple Loops, 148
    Audio and MIDI, 146
    From Browsers, 146
    From the Desktop, 146
    Opening the All Files Browser, 147
    Opening the Media Browser, 146
    Supported Audio Files, 146
    Using Import Command, 147
    Using the Import Command, 147, 148
Input Monitoring
    Enabling, 141
Inserts
    Plug-Ins, 190
Installation Requirements, 8

J–K

Keyboard Shortcuts, 22, 264
    Bounce and Export, 267
    Edit Commands, 266
    Main Window, 264
    Navigation and Playback, 265
    Other Windows, 265
    Tracks, 266
    Zoom Functions, 265

L

Latch Automation Mode, 221

Launching Applications, 17
    From the Desktop, 18
    Mac-Based Systems, 17
    Windows-Based Systems, 18
LCD, 110
    Playhead Position, 110
Limiter, 200, 202, 228
    Usage, 228
Logic Pro X
    Capabilities, 40
    Hardware Options, 41
    Included Options, 41
    Installation, 41, 42
    Launching, 44
    System Requirements, 42

M

Mac Versus Windows, 3
    Mac-Based Computers, 3
    Running Windows on a Mac, 4
    Windows-Based Computers, 3
Main Window, 105
    Horizontal Scrolling, 123
    Keyboard Shortcuts, 264
    Scrolling, 122
    Transport Controls, 110
    Vertical Scrolling, 123
    Zooming, 122
Marquee tool, 120
Media Browser, 146
Merge Takes
    Enabling, 145
Metering, 169
    Output Channel Strip, 170
Metronome Click, 138
Microphone Accessories
    Acoustic Shields, 63
    Connections, 64
    Mike Stand, 63
    Pop Filters, 63
    Shock Mounts, 63
    Vocal Booths, 63
    Windscreens, 63
Microphones, 59
    Accessories, 63
    Basic Setup, 64
    Basic Techniques, 62
    Condenser, 60
    Connections, 64
    Dynamic, 59
    Polar Pattern, 61
        Bidirectional, 62
        Cardiod, 62
        Omnidirectional, 62
    Position, 64
    Proximity Effect, 62
    Ribbon, 60
    USB, 60
MIDI, 78
    Automation/MIDI area, 253
    Cables and Jacks, 89
    Controller Messages, 82
    Creating Notes, 254
    Editing, 254
    Editing Techniques, 251
    Event List, 81
    History, 78
    Interfaces, 90
    MIDI versus Audio, 91
    Monitoring, 92
    Multiple Devices, 91
    Note Off, 81
    Note On, 81
    Note Values, 80
    Notes, 80
    Piano Roll Editor, 253
    Plug-and-Play Setup, 89
    Program Changes, 81
    Protocol, 80
    Setup and Signal Flow, 89
    Track Views, 251

Velocity, 80, 258
What is MIDI?, 91
Working with Regions, 252
MIDI Controllers, 83
Alternative Controllers, 87
Digital Pianos, 83
Drumpad Controllers, 86
Grid Controllers, 86
Keyboard Controllers, 85
Purchasing, 88
Synthesizers, 83
What to Look for, 87
MIDI Editing
Manual Note Entry, 254
Selecting Groups of Notes, 255
Selecting Indvidual Notes, 255
Selecting Notes, 255
MIDI Event List, 81
MIDI Notes
Deleting, 256
Editing, 256
Moving, 256
Transposing, 257
Trimming, 257
MIDI Regions
Editing, 154
MIDI Velocity
Editing, 258
Mixdown, 232
Creating a Mixdown, 228
Options, 232
Mixer
Record and Playback Controls, 116
Signal Routing Controls, 116
Mixer Display, 115
Mixer Window, 115
Mixing
Basic, 168
Clipping, 169
In the Box, 178
Metering Channel Strips, 169

Output Channel Strip Metering, 170
Panning, 172
Setting Levels, 168, 171
Submixing, 261
Using Mute, 172
Using Solo, 171
Mixing in the Box, 178
Advantages, 178
Monitoring, 140
Bounce Progress, 234
MIDI Controller, 143
Moog, Bob, 78
Multipressor, 202
Multi-Tracking, 64
Recording, 65
Signal Flow, 64, 66
Mute
Using, 172

N

Navigation
Keyboard Shortcuts, 265
Noise Gate, 202
Noise Shaping, 231
Non-Destructive Editing, 154

O

Options
Bouncing, 232
Keyboards, 10
Keypads, 10
Mice and Trackballs, 10
Onboard Sound, 10
Trackpads and Touchpads, 10
Output Channel Strip
Metering, 170

P

Panning, 172
    Mono Tracks, 172
    Stereo Tracks, 173
Pencil tool, 120
Piano Roll, 260
Playback
    Keyboard Shortcuts, 265
Playback Behavior, 124
    Setting a Location, 124
Playhead
    Monitoring, 125
Playhead Location
    LCD, 125
Playhead Position, 110
Plug-In Parameters, 195
    Adjusting with Keyboard, 196
    Adjusting with Mouse, 195
    Fine Adjustment, 195
    Return to Default, 196
Plug-Ins
    AAX, 39
    Adjusting Parameters, 195
    Audio Units, 39
    AudioSuite, 40
    Basics, 190
    Bypass Control, 193, 195
    Close Control, 195
    Common Controls, 192
    Delay, 204
    Displaying Plug-In Windows, 192
    Duplicating, 192
    Dynamics, 199
    Effects, 38
    EQ, 196
    Formats, 38, 39
    Inserting, 190
    Link Control, 194
    Moving, 192
    Removing, 191
    Replacing, 191
    Reverb, 204
    RTAS, 40
    Settings Controls, 193
    TDM, 40
    Time-Based, 204
    View Control, 194
    Virtual Instrument, 38
    VST, 39
Pointer tool, 120
POW-r Dither
    Noise Shaping, 231
Pro Tools Systems, 40
Processing, 176
    External Gear, 177
    Gain-Based, 176
    Inserts vs Sends, 177
    Stereo Output, 228
    Time-Based Effects, 177
Project Chooser, 102

Q–R

Quantization, 246, 260
    Scale Quantize, 261
    Time Quantize, 260
RAM, 3, 4, 5
    Installing More, 5
    Requirements, 5
Read Automation Mode, 221
Recalling a Saved Mix, 220
Record and Playback Controls, 116
Record-Enabling Tracks, 140
Recording
    Beginning Immediately, 141
    Initiating a Take, 141
    Making a Selection, 139
    Metronome Click, 138
    Monitoring, 140
    Multiple Takes, 142
    Record-Enabling Tracks, 140

Setting Tempo, 138
Setting Up, 138
Signal Flow, 66
Stop and Delete Region, 142
Stopping a Take, 141
Stopping Immediately, 141
Recording Audio, 139
Recording MIDI, 143
Merge Takes, 145
Monitoring a MIDI Controller, 143
Record-Enabling Tracks, 144
region colors, 176
Region Inspector, 259
View Settings, 259
Regions
Copying, 153
Cutting, 153
Cutting, Copying, and Pasting, 152
Moving, 152, 153
Pasting, 153
Selecting, 152
Selecting Entire Regions, 152
Selecting Partial Regions, 152
Repeat Command, 252
Repeat Multiple Command, 252
Reverb, 204
Applications, 205
Definition, 204
Usage, 205
Rulers, 114

S

Saving Files, 18
Scissors tool, 120
Scrolling, 122
Scrolling Options, 124
Page Scrolling, 125
Selections, 126
Editing, 128
Playback, 128

Recording, 128
Send-and-Return, 209
Signal Flow, 64
Signal Routing Controls, 116
Signals
Wet vs Dry, 208
Snap to Grid, 149
Setting the Snap Value, 149
Software Instrument
Region Inspector, 259
Solo
Using Solos, 171
Stems
Creating Stems, 263
Stereo Output
Limiting, 228
Storage, 7, 9
External Drives, 9
Hard Disk Drive, 7
HDD Versus SSD, 9
Solid State Drive, 7
Submix, 261
Creating a Track Stack, 262

T

Take Folder
Enabling, 145
Tempo
Setting Tempo, 138
Text tool, 120
Time-Based Effects, 177
As Inserts, 208
Send-And-Return, 209
Timeline Location
Ruler Displays, 125
Tools
Selecting, 151
Touch Automation Mode, 221, 222
track colors, 175

Track Stacks, 261
    Creating, 262
Tracks
    Adding, 117
    External MIDI vs Software Instrument, 118
    Keyboard Shortcuts, 266
    Mono vs Stereo, 118
transient markers, 247
Transport Controls, 121
    Fast Forward, 121
    Go To Beginning, 121
    Play, 121
    Record, 122
    Rewind, 121
    Stop, 121
Transport Window, 110
    Controlling Plaback, 124
    Controls, 121
Trimming
    Region End, 155
    Region Start, 154

U–V

Velocity, 258
Vertical Scrolling, 123
Virtual Instruments
    Assigning, 93
    Tracking, 93

W

Warping Audio, 247
Wet vs Dry Signals, 208
Working with an Application, 19
    Keyboard Shortcuts, 22
    Using Menus, 20
Write Automation Mode, 221

X, Y, Z

Zoom tool, 120
Zooming, 122
    Alternate Functions, 123
    Keyboard Shortcuts, 265

About the Authors

This book was written by Ryan Rey and Harry Gold and published under the direction of NextPoint Training, Inc. The course is designed to prepare students for certification in basic audio production using Logic Pro X software under the NextPoint Training Digital Media Production program.

Ryan Rey is a composer, guitarist, educator, and San Francisco Bay Area native. He teaches courses in Logic Pro X, music theory, piano, and sound for picture at Pyramind in San Francisco; studio recording and AV technology at Diablo Valley College in Pleasant Hill; and a variety of subjects online through Pryamind's Mentorship Network.

Active in various non-profit organizations, Rey is currently Director of the concert series Seventh Avenue Performances in San Francisco. He is also on the board of directors for Ninth Planet, an organization that promotes and presents new music by living composers, formerly known as Composers, Inc., of which he was the former Executive Director and Artistic Co-Director. Rey played guitar in the death metal band Antagony and the chiptune/folk/doom/chamber music group The Mineral Kingdom. Recent activities include making video tips and tutorials on music theory and audio production, and composing for production music library CrimeSonics.

Harry Gold is a Bay Area-based blues and jazz guitarist, pianist, arranger, and singer who has performed all over the United States and Europe. He is also an audio engineer, long-time Pro Tools user, and Avid Certified Instructor. Currently, Gold works as an Adjunct Professor at Academy of Art University, and as a part-time instructor at Pyramind Inc. in San Francisco. He teaches classes in Pro Tools, Logic, Score Preparation, Arranging and Composition.

Besides performing and teaching, Gold has worked as a luthier, repairing and building guitars for over a decade. He also runs Fools Gold Studios, a recording studio based in Berkeley, California specializing in live recording.

Previously, Gold developed and directed the music program at Holden High School in Orinda, California. He started the school's first audio engineering program and was able to certify students as Pro Tools Operators in music and post-production. Gold holds a Master's of Music from the Berklee College of Music, and a Bachelor's of Music from the California Jazz Conservatory.